AF753673

FOOD SCIENCE AND TECHNOLOGY

GENETICALLY ENGINEERED SALMON

BACKGROUND AND ISSUES

FOOD SCIENCE AND TECHNOLOGY

Additional books in this series can be found on Nova's website under the Series tab.

Additional e-books in this series can be found on Nova's website under the e-book tab.

FOOD SCIENCE AND TECHNOLOGY

GENETICALLY ENGINEERED SALMON

BACKGROUND AND ISSUES

DIEDRA P. THARP
EDITOR

nova publishers
New York

Copyright © 2014 by Nova Science Publishers, Inc.

All rights reserved. No part of this book may be reproduced, stored in a retrieval system or transmitted in any form or by any means: electronic, electrostatic, magnetic, tape, mechanical photocopying, recording or otherwise without the written permission of the Publisher.

For permission to use material from this book please contact us:
Telephone 631-231-7269; Fax 631-231-8175
Web Site: http://www.novapublishers.com

NOTICE TO THE READER

The Publisher has taken reasonable care in the preparation of this book, but makes no expressed or implied warranty of any kind and assumes no responsibility for any errors or omissions. No liability is assumed for incidental or consequential damages in connection with or arising out of information contained in this book. The Publisher shall not be liable for any special, consequential, or exemplary damages resulting, in whole or in part, from the readers' use of, or reliance upon, this material. Any parts of this book based on government reports are so indicated and copyright is claimed for those parts to the extent applicable to compilations of such works.

Independent verification should be sought for any data, advice or recommendations contained in this book. In addition, no responsibility is assumed by the publisher for any injury and/or damage to persons or property arising from any methods, products, instructions, ideas or otherwise contained in this publication.

This publication is designed to provide accurate and authoritative information with regard to the subject matter covered herein. It is sold with the clear understanding that the Publisher is not engaged in rendering legal or any other professional services. If legal or any other expert assistance is required, the services of a competent person should be sought. FROM A DECLARATION OF PARTICIPANTS JOINTLY ADOPTED BY A COMMITTEE OF THE AMERICAN BAR ASSOCIATION AND A COMMITTEE OF PUBLISHERS.

Additional color graphics may be available in the e-book version of this book.

Library of Congress Cataloging-in-Publication Data

ISBN: 978-1-61122-558-7

Published by Nova Science Publishers, Inc. † New York

CONTENTS

PREFACE

This book discusses the genetically engineered salmon; the background of the techniques and the issues that go along with it.

Chapter 1 – If approved by the Food and Drug Administration (FDA), Atlantic salmon would be the first genetically engineered (GE) animal to be marketed in the United States for human consumption. Genetic engineering techniques are used by scientists to insert genetic material from one organism into the genome of another organism. Genetically engineered salmon have been modified to grow more quickly and use feed more efficiently. However, some are concerned that, in this rapidly evolving field, current technological and regulatory safeguards are inadequate to protect the environment and ensure that these products are safe to be used as food.

Over seventeen years ago, AquaBounty Technologies Inc. first applied to the FDA for approval of a genetically engineered Atlantic salmon. In 2009, AquaBounty submitted to the FDA the last required study for their new animal drug (NAD) application. The FDA is regulating GE Atlantic salmon as an NAD under the Federal Food, Drug, and Cosmetic Act (FFDCA; 21 U.S.C. §321). An NAD is approved by the agency only after the drug is shown to be safe and effective. On September 19-21, 2010, the FDA's Veterinary Medicine Advisory Committee (VMAC) met to consider the AquaBounty application for approval of GE salmon, trademarked as AquAdvantage salmon, for human consumption, and held a public hearing. On December 20, 2012, FDA announced the availability for public comment of (1) a draft environmental assessment of the proposed conditions specified by AquaBounty and (2) FDA's preliminary finding of no significant impact (FONSI) for AquaBounty's conditions. In the FONSI, FDA reiterated that food from AquAdvantage salmon is as safe as food from non-GE salmon and determined that there are no significant food safety hazards or risks associated with AquAdvantage salmon. FDA may still require a full environmental impact statement (EIS) prior to approval of AquaBounty's application. No further action has been taken by FDA, and the Aquabounty NAD application is still under consideration.

Environmental concerns related to the development of GE salmon include the potential for competition and interbreeding with wild fish. According to some, escaped GE salmon could spawn with wild Atlantic salmon and introduce the modified genetic material to the wild population. Sterilization and bioconfinement have been proposed as a means of isolating GE salmon to minimize the likelihood of harm to wild fish populations. To address these concerns, AquaBounty proposes to produce salmon eggs (all sterile females) in Canada, ship these eggs to Panama, grow and process fish in Panama, and ship table-ready, processed fish to the United States for retail sale. Additional concerns have been voiced concerning food safety, labeling of GE salmon, and economic effects on existing wild salmon fisheries.

Some have asserted that FDA approval of AquAdvantage salmon is overdue and that delays have hindered investment and development of the U.S. biotechnology sector. Others have questioned the adequacy of the FDA's review of GE salmon and whether the existing approval process is equipped to fully evaluate the risks of this technology, especially potential environmental harm. In response to food safety and environmental concerns, legislation has been introduced during the 113th Congress, including S. 246 and H.R. 1667, which would prohibit the transport, sale, possession, release, or use of GE fish. H.R. 584 and S. 248 would amend Section 403 of the FFDCA by adding a requirement to label genetically engineered fish, while H.R. 1699 and S. 809 would require labeling of foods produced using genetic engineering, including fish. No further action has been taken on these or other bills which would require additional regulation of genetically engineered organisms.

Chapter 2 – Report of the Veterinary Medicine Advisory Committee on Briefing Packet AquAdvantage Salmon.

In: Genetically Engineered Salmon: Background and Issues ISBN: 978-1-61122-558-7
Editor: Diedra P. Tharp © 2014 Nova Science Publishers, Inc.

Chapter 1

GENETICALLY ENGINEERED SALMON*

Harold F. Upton and Tadlock Cowan

SUMMARY

If approved by the Food and Drug Administration (FDA), Atlantic salmon would be the first genetically engineered (GE) animal to be marketed in the United States for human consumption. Genetic engineering techniques are used by scientists to insert genetic material from one organism into the genome of another organism. Genetically engineered salmon have been modified to grow more quickly and use feed more efficiently. However, some are concerned that, in this rapidly evolving field, current technological and regulatory safeguards are inadequate to protect the environment and ensure that these products are safe to be used as food.

Over seventeen years ago, AquaBounty Technologies Inc. first applied to the FDA for approval of a genetically engineered Atlantic salmon. In 2009, AquaBounty submitted to the FDA the last required study for their new animal drug (NAD) application. The FDA is regulating GE Atlantic salmon as an NAD under the Federal Food, Drug, and Cosmetic Act (FFDCA; 21 U.S.C. §321). An NAD is approved by the agency only after the drug is shown to be safe and effective. On September 19-21, 2010, the FDA's Veterinary Medicine Advisory Committee (VMAC) met to consider the AquaBounty application for approval of GE salmon, trademarked as AquAdvantage salmon, for human consumption, and held a public hearing. On December 20, 2012, FDA announced the availability for public comment of (1) a draft environmental assessment of the proposed conditions specified by AquaBounty and (2) FDA's preliminary finding of no significant impact (FONSI) for AquaBounty's conditions. In the FONSI, FDA reiterated that food from AquAdvantage salmon is as safe as food from non-GE salmon and determined that there are no significant food safety hazards or risks associated with AquAdvantage salmon. FDA may still require a full environmental impact statement (EIS) prior to approval of AquaBounty's application. No further action has been taken by FDA, and the Aquabounty NAD application is still under consideration.

* This is an edited, reformatted and augmented version of a Congressional Research Service publication R43518, prepared for Members and Committees of Congress dated April 30, 2014.

Environmental concerns related to the development of GE salmon include the potential for competition and interbreeding with wild fish. According to some, escaped GE salmon could spawn with wild Atlantic salmon and introduce the modified genetic material to the wild population. Sterilization and bioconfinement have been proposed as a means of isolating GE salmon to minimize the likelihood of harm to wild fish populations. To address these concerns, AquaBounty proposes to produce salmon eggs (all sterile females) in Canada, ship these eggs to Panama, grow and process fish in Panama, and ship table-ready, processed fish to the United States for retail sale. Additional concerns have been voiced concerning food safety, labeling of GE salmon, and economic effects on existing wild salmon fisheries.

Some have asserted that FDA approval of AquAdvantage salmon is overdue and that delays have hindered investment and development of the U.S. biotechnology sector. Others have questioned the adequacy of the FDA's review of GE salmon and whether the existing approval process is equipped to fully evaluate the risks of this technology, especially potential environmental harm. In response to food safety and environmental concerns, legislation has been introduced during the 113th Congress, including S. 246 and H.R. 1667, which would prohibit the transport, sale, possession, release, or use of GE fish. H.R. 584 and S. 248 would amend Section 403 of the FFDCA by adding a requirement to label genetically engineered fish, while H.R. 1699 and S. 809 would require labeling of foods produced using genetic engineering, including fish. No further action has been taken on these or other bills which would require additional regulation of genetically engineered organisms.

INTRODUCTION

By a broad definition, "genetic modification" refers to changes in an organism's genetic makeup that do not occur in nature. For millennia, farmers and scientists have modified the genetics of animals by selecting those individuals with desirable traits for further breeding. With the advent of modern biotechnology (e.g., genetic engineering or bioengineering), it is now possible to take a gene (or genes) for a specific trait from an organism and transfer it to another organism of a different species. For the purpose of FDA's guidance to industry, FDA defines genetically engineered (GE) animals as those animals modified by recombinant DNA (rDNA)[1] techniques, including the entire lineage of animals that contain the modification.[2]

Recombinant DNA techniques expand the range of traits that may be transferred to another organism and increase the speed and efficiency by which desirable traits may be incorporated into organisms. Desirable traits may reduce production costs and sometimes make the organism or products made from it more desirable to consumers. Genetically engineered plant varieties, such as herbicide-resistant corn and soybeans, have already been widely adopted by U.S. farmers. These techniques are now being used to develop genetically engineered organisms for the aquaculture industry.

Approximately 50 species of fish have been subject to genetic modification and more than 400 fish/trait combinations have been developed.[3] Fish and other marine organisms are being modified to reduce production costs of human food, to produce pharmaceuticals, to test water contamination, and for other uses.[4] Fish are particularly attractive candidates for genetic engineering because they produce eggs in large quantities and their eggs are more easily manipulated because they are fertilized and develop externally. Aquaculture also supplies a rapidly expanding market of different seafood products. Countries with active research programs for genetically engineered fish include China, Cuba, India, Korea, the Philippines, and Thailand.

Development of GE fish has prompted some advocacy groups to raise a number of environmental concerns.[5] If fish are accidently released into the environment they may spread quickly and be difficult to contain. GE fish that escape to the wild could compete with wild fish and harm wild populations. Another concern is that GE fish may interbreed with wild fish and allow the modified genetic material to become assimilated into the wild fish population.[6] Sterilization and bioconfinement have been proposed as a means of isolating GE fish to minimize the potential for interactions with wild fish populations and changes to related ecosystems.

Food safety concerns also have been expressed by consumer groups who question whether genetically modified organisms could pose unique hazards to public health.[7] Furthermore, some in the fishing industry are concerned that greater efficiency in the aquaculture industry could harm salmon fisheries.[8] Salmon farm production in the 1990s depressed salmon prices and affected fishing businesses and coastal communities that depend on wild fisheries.

Some have asserted that FDA approval of GE salmon is overdue and has been delayed due to political interference.[9] They conclude that delays have hindered investment and development of the U.S. biotechnology sector. They also question whether biotechnology industries in the United States will be able to compete with companies in other parts of the world. Moreover, they also infer that the availability of GE salmon could benefit consumers who are seeking low-fat and affordable options.[10]

During the 112th and 113th Congresses, the adequacy of the FDA's review of GE animals has been questioned. Several bills introduced in the 112th Congress would have amended the Federal Food, Drug, and Cosmetic Act (FFDCA) to prevent approval of GE salmon or to require labeling of genetically engineered fish. None of these bills was enacted. Several bills have been introduced during the 113th Congress, including S. 246 and H.R. 1667, which would prohibit the possession or use of GE fish in the United States. H.R. 584, H.R. 1699, S. 248, and S. 809 would add a requirement under the FFDCA to label genetically engineered fish. No further action has been taken on these or other bills that would mandate additional regulation of genetically engineered organisms.

U.S. Biotechnology Regulation and Oversight

Coordinated Framework for Regulation of Biotechnology

Federal guidance for regulating biotechnology products is provided in the Coordinated Framework for Regulation of Biotechnology (51 *Fed. Reg.* 23302), published in 1986 by the White House Office of Science and Technology Policy (OSTP). A key regulatory principle of the U.S. biotechnology regulatory structure is that genetically engineered products should continue to be regulated according to their characteristics and unique features, not their production method— that is, whether or not they were created through biotechnology. The framework provides a regulatory approach intended to ensure the safety of biotechnology research and products, using existing statutory authority and previous agency experience with traditional breeding techniques. The three lead agencies are USDA's Animal and Plant Health Inspection Service (APHIS), the Food and Drug Administration (FDA) at the Department of Health and Human Services, and the Environmental Protection Agency (EPA).

In 2002, the National Research Council (NRC) published its report on animal biotechnology.[11] Some newer applications of biotechnology did not exist when the current regulatory framework was enunciated. The NRC animal biotechnology report concluded that this General Framework "might not be adequate to address unique problems and characteristics associated with animal biotechnologies" and that federal agency responsibilities are not clear.[12]

FDA Regulatory Framework

FDA regulates food, animal feed additives, and human and animal drugs, including those from biotechnology, primarily to ensure that they pose no human health risks, mainly under the Federal Food, Drug, and Cosmetic Act (21 U.S.C. §301 et seq.) and the Public Health Service Act (42 U.S.C. §201 et seq.). Under the FFDCA, all food and feed manufacturers must ensure that the domestic and imported products they market, except for most meats and poultry, are safe and properly labeled, including those developed through genetic engineering.

FDA has stated that most—although probably not all—gene-based modifications of animals for production or therapeutic claims fall within the purview of the agency's Center for Veterinary Medicine (CVM), which regulates them under the FFDCA as new animal drugs (NAD) (21 U.S.C. §321).[13] Under the FFDCA, drugs are defined in Section 201(g) as "articles intended for the use in the diagnosis, cure, mitigation, treatment, or prevention of disease in man or other animals; and articles (other than food) intended to affect the structure or any function in the body of man or other animals."[14]

On January 15, 2009, CVM released its industry guidance for producers and developers of GE animals and their products. The guidance provides an approach to satisfy applicable statutes and regulations.[15] The guidance states (on page 6): "The rDNA construct in a GE animal that is intended to affect the structure or function of the body of the GE animal, regardless of the intended use of products that may be produced by the GE animal, meets the FFDCA drug definition."

A new animal drug (NAD) is assumed to be "unsafe" unless FDA has approved a new animal drug application (NADA) for that particular use, or the NAD is for investigational use and subject to an exemption from the drug approval requirement (among a few other specified exemptions). FFDCA and federal regulations describe information that must be submitted to FDA as part of NADAs.[16] The industry guidance lays out the pre-market approval process, including the information required of applicants which fulfills the regulatory requirements. Required information is broken out into six categories, which include product identification, molecular characterization of the construct, molecular characterization of the GE animal lineage, phenotypic characterization of GE animal, genotypic and phenotypic durability assessment, and the food/feed safety and environmental safety assessments.

The food safety assessment includes examination of both the direct toxicity (including allergenicity) potential of food from a GE animal as well as any indirect toxicity. Food and feed will be considered safe if the composition of edible materials from the GE animal can be shown to be "substantially equivalent" to that from a non-GE animal. Therefore, if animals of the same or comparable type are commonly and safely consumed, there is a presumption that food from the GE animal is safe[17] and the product will not have to be labeled.

An FDA decision regarding an NADA is a federal action subject to the National Environmental Policy Act (NEPA; 42 U.S.C. 4321).[18] NEPA requires federal agencies to consider the environmental consequences of an action *before* proceeding with it and to involve the public in its decision-making process. To demonstrate compliance with NEPA, federal agencies must prepare an environmental impact statement (EIS) for federal actions anticipated to have "significant" impacts on the environment. The EIS is a detailed evaluation of the proposed action and provides opportunity for the public, other federal agencies, and outside parties to provide input into the process. In this case, the FDA is ultimately responsible for determining whether or not an EIS is necessary.[19]

To assist the agency in making the determination of whether an EIS is necessary, an applicant for an NAD must submit documentation to support a claim for a categorical exclusion or for drafting a preliminary environmental assessment (EA). Actions that, based on an agency's past experience with similar actions, have no significant impacts are categorically excluded from the requirement to prepare an EA or EIS. If an EA is developed and it is found that the NAD would have no significant environmental impacts, the agency would issue a finding of no significant impact (FONSI). If the EA determines that the environmental consequences of an NAD are anticipated to be significant, an EIS is prepared. In cases where significant impacts are anticipated, the federal agency may decide to prepare an EIS without first preparing an EA.

Under the NAD regulatory protocols, FDA must keep all information about a pending drug application confidential, with the exception of information publicly disclosed by the manufacturer, to protect proprietary information. This approach can limit the opportunity for public comment before approval. Given that the AquaBounty salmon could be the first GE animal approved for human consumption, some critics have called for more transparency during the pre-market approval process.[20]

The FDA guidance does not require evaluation of genetically engineered organisms such as GE salmon as a food additive. A food additive is "any substance, the intended use of which results directly or indirectly, in it becoming a component or otherwise affecting the characteristics of food."[21] Food additives require pre-market approval from FDA, unless the additive has been evaluated by scientific experts and determined to be "generally regarded as safe" (GRAS). If made subject to a food additive review, products such as GE salmon would have to undergo comprehensive toxicological studies. Critics have questioned whether the NAD regulatory review is sufficient and have petitioned for evaluation of GE salmon under FDA's food additive requirements.

U.S. Department of Agriculture (USDA)

Several USDA agencies, operating under a number of statutory authorities, also have at least potential roles in the regulation of transgenic and cloned animals and their products. As several critical reviews have indicated, USDA has not had a clearly spelled out policy in this area, including whether it intends to exercise these authorities to regulate GE animals.[22] USDA's Animal and Plant Health Inspection Service earlier had expressed its intention to publish an advance notice of proposed rulemaking (ANPR) on GE animals, possibly in 2008.[23] Instead, in concert with FDA's notice on its draft guidance, APHIS published, in the September 19, 2008, *Federal Register*, a request for information from the public and scientists on how GE animals might affect U.S. animal health.[24] Over 670 comments were received by November 18, 2008, as they had been for the FDA draft guidance. Most of the comments were outside APHIS's authority under the Animal Health Protection Act. FDA issued its final guidance for developers of GE animals on January 15, 2009. The guidance states that FDA intends to develop a memorandum of understanding with APHIS to determine its role in the comprehensive oversight of GE animals.

APHIS has broad authority, under the Animal Health Protection Act (AHPA; 7 U.S.C. §8301 et seq.), to regulate animals and their movement to control the spread of diseases and pests to farm-raised animals. APHIS also administers the Viruses, Serums, Toxins, Antitoxins, and Analogous Products Act (21 U.S.C. §151-159), aimed at assuring the safety and effectiveness of animal vaccines and other biological products, including those of GM origin, and the Animal Welfare Act (7 U.S.C. §2131 et seq.), portions of which govern the humane treatment of several kinds of warm-blooded animals used in research (but generally not agricultural animals or cold-blooded animals such as fish). Elsewhere at USDA, the Food Safety and Inspection Service (FSIS) is responsible for ensuring the safety and proper labeling of most food animals and meat and related products derived from them under the Federal Meat Inspection Act (21 U.S.C. §601 et seq.) and Poultry Products Inspection Act (21 U.S.C. §451 et seq.).

Labeling of Food Containing GE Material

Federal food labeling policy, including the labeling of foods containing bioengineered material, is regulated under the Federal Food, Drug, and Cosmetic Act of 1938 (21 U.S.C. §§301 et seq.) and the Fair Packaging and Labeling Act of 1966 (P.L. 89-755;15 U.S.C. §§1451 et seq.).[25] Section 403 of the FFDCA governs food labeling. Under Section 403(a)(1), a food is considered misbranded if its labeling is false or misleading. Section 201(n) of the FFDCA provides additional guidance on how food labeling may be misleading. It states that a label is misleading if it fails to reveal facts that are material in light of representations made or suggested in the labeling, or material with respect to consequences that may result from the use of the food prescribed in the labeling, or under such conditions of use that are customary or usual.

The United States does not require mandatory labeling that identifies foods containing genetically modified material.[26] The notion of "substantial equivalence" guides food labeling requirements; if a food containing GE material is "substantially equivalent" to a food not containing GE material, federal regulations do not require that it be labeled as containing GE material. When there is no material difference between products, FDA does not have the authority to require labeling on the basis of consumer interest alone. If there is a material difference between GE and non-GE foods, FDA could require such differences to be identified in food labeling.[27] Companies that might wish to label their foods as not containing bioengineered products may do so, if they can definitively show that the foods do not contain GE products.

Genetically engineered crops became commercially available in the mid-1990s.[28] Generally, FDA has not found that food from GE organisms warrants different or greater safety concerns than non-GE organisms or exhibits different characteristics such as nutritional value or functional characteristics than from non-GE organisms. Today, oil from bioengineered soy and canola, soy protein, and high fructose corn syrup can be found in many manufactured foods, perhaps as high as 60%-70% of processed foods. The FDA has found most GE crops to be "substantially equivalent" to non-GE crops and approved their safety for human consumption in processed foods.

In 1992, FDA published a policy statement on foods derived from new plant varieties, including those developed through genetic engineering.[29] This policy statement did not establish any special labeling requirements for bioengineered foods as a particular class of foods. FDA stated that the agency had no basis for concluding that bioengineered foods differed in any meaningful way from non-bioengineered foods, and therefore had no basis for requiring that such foods be labeled.[30]

Although the 1992 policy statement did not require special labeling for bioengineered foods, FDA did advise that any labeling requirements that apply to foods in general also apply to bioengineered foods. Under Section 201(n), the label of the food must reveal all "material facts" about the food. FDA currently requires special labeling of bioengineered foods if the food has a significantly different nutritional property. For example, if a new food contained an allergen that consumers would not expect to be present, or if a food contained a toxic ingredient above acceptable limits, FDA would require that the food be labeled as such.

FDA subsequently issued a 2001 draft guidance document for the voluntary labeling of foods that have or have not been developed by bioengineering. These guidelines address how the agency interprets revealing "material facts" about a food on a label. In their guidelines, FDA suggests that terms such as "GMO free" or "not genetically modified" could be technically inaccurate and misleading. On the other hand, labeling statements that the food or its ingredients were not created by bioengineering processes would likely be appropriate.[31] For example, currently, no bioengineered watermelons are on the market. A statement that a watermelon was not genetically engineered might be deemed misleading by FDA because it implies that other watermelons might be bioengineered.

General Mandatory Labeling Issues

Mandatory labeling of bioengineered products in the United States has been proposed at the national, state, and local levels. No labeling requirement has been enacted. Proponents of mandatory labeling for bioengineered foods argue that consumers should have the right to know what they are purchasing. Even if the FDA states that a bioengineered product is, from a food safety perspective, "substantially equivalent" to its traditional counterpart, labeling proponents assert that the consumer should be able to choose between those foods that may contain bioengineered products and those that do not. Some proponents of labeling also argue that for religious or ethnic reasons, many consumers may want to avoid eating animal products, including processed food products that contain animal genetic material.

With the widespread adoption of bioengineered plants and their now ubiquitous use in food processing, labeling opponents point to the logistical difficulties and costs of ensuring a food product does not contain bioengineered ingredients. As the global food system is currently constructed, segregating bioengineered products from non-bioengineered products would be technically complex and costly.[32] Labeling opponents also argue that the increased food prices as a result of labeling would be borne by all consumers, not just those who wish to avoid bioengineered products. Opponents also point out that, under the regulatory basis of "substantial equivalence," to label a bioengineered food product as such would suggest to a consumer that something is unhealthy about the bioengineered food product, and would be prejudicial.

GE Salmon Background

Wild Salmon Production

Salmon is a general name for anadromous species which belong to the taxonomic family known as *salmonidae*. Anadromous species live in fresh water during early stages of their life (egg, fry, and juvenile), migrate to the ocean to grow to adult size, and when sexually mature, return to freshwater to spawn. Most salmon production in the United States is from the wild and consists of five main species of Pacific salmon which include Chinook salmon *Onchorhynchus tshawytscha*; sockeye salmon *Onchorhynchus nerka*; coho salmon *Onchorhynchus kisutch*; pink salmon *Onchorhynchus gorbuscha*; and chum salmon *Onchorhynchus keta*.

Over 95% of wild commercial salmon production is from Alaskan stocks. These stocks are abundant and productive, largely because there have been relatively few human disturbances on major salmon rivers in Alaska such as dams. Stocks in the Pacific Northwest vary widely in health, but most are in relatively poor condition because of historic overfishing and degradation of riverine habitat. These stocks are of limited commercial importance although they still support recreational fisheries and contribute to local economies.[33] Production in both Alaska and the Pacific Northwest is enhanced by releases of salmon from hatcheries which are subsequently harvested after growing to adult size in the ocean. In 2012, total commercial landings of Pacific salmon were 288,400 metric tons with a value of $489.1 million.[34]

There are no commercial wild fisheries for Atlantic salmon in North America, and only limited numbers of Atlantic salmon *Salmo salar* spawn in New England rivers. Many Atlantic salmon runs are at historically low levels, especially in the southern parts of their range, such as New England. Population declines have been caused by overfishing, dams, and degraded environmental quality of rivers. The remainingAtlantic salmon runs in Maine are listed as endangered under the Endangered Species Act (16 U.S.C. §§1531-1543). A limited recreational fishery exists in Canada and small commercial fisheries remain in the North Atlantic.

Salmon Aquaculture

For the aquaculture industry, salmon is a desirable candidate for genetic engineering because of high consumer demand for salmon products. Salmon aquaculture technology is well developed and commercial salmon farming has been established in many temperate countries. Of the salmon species used for aquaculture, Atlantic salmon account for most production. Atlantic salmon grow well under culture conditions and adapt well to culture conditions outside its range.[35] Most production is from net pens that are suspended in coastal waters, but salmon can be grown in freshwater raceways, tanks, or recirculating systems where adequate water supplies are available. Salmon farming has an advantage over the wild seasonal fishery because it can provide a consistent fresh product throughout the year.

During the 1970s intensive commercial salmon farming in net pens was adopted in Norway and production expanded rapidly. Salmon production in other countries with suitable coastal areas such as Great Britain, Chile, and Canada followed and also increased rapidly. Production costs decreased with improvements in broodstock quality, feed, disease management, and other production factors. During the period of rapid expansion of salmon farming, prices for both cultured and wild salmon have generally trended downward. Environmental concerns also emerged because of potential harm to wild fish stocks used for salmon feed, possible transfer of disease from farmed to wild salmon stocks, environmental effects of fish wastes and lost feed from open water cages, transfer of therapeutic agents used for cultured fish to the environment, and escapes of cultured fish.[36] The aquaculture salmon industry has reportedly made progress in addressing some of these concerns and has begun adopting best management practices to decrease external effects on the environment.[37]

Production and Trade

In 1996, world-wide salmon farm production exceeded commercial harvest of wild salmon,[38] and in 2011, aquaculture production of salmon, trouts, and smelts was 2.773 million metric tons (mmt) with a value of $15.174 billion.[39] Production of Atlantic salmon, the main species of this group, was 1.721 mmt with a value of $9.710 billion.[40] Norway led production followed by Chile and the United Kingdom (**Table 1**). In 2011, farmed production of Atlantic salmon in the United States was 18,595 metric tons with a value of $104 million.

Global trade of salmon products has continued to increase with gains in farmed production. Although the United States is a major producer and consumer of wild Pacific salmon, in 2011, U.S. imports of fresh and frozen farmed Atlantic salmon totaled 181.2 thousand metric tons with a value $1.520 billion. During the last decade, Atlantic salmon imports have accounted for approximately 50% to 60% of the total U.S. fresh and frozen salmon supply.[41]

Table 1. Atlantic Salmon Farm Production and U.S. Imports of Atlantic Salmon

Country	2011 Atlantic Salmon Production (metric tons)	2011 Atlantic Salmon Value (US Dollars 000s)	2011 Exports to the United States (metric tons)	2011 Exports to the United States (US Dollars 000s)
Norway	1,059,958	4,855,561	18,850	222,054
Chile[a]	264,319	2,223,175	54,607	563,398
United Kingdom	158,018	942,642	15,863	113,566
Canada	102,064	613,523	70,995	478,342

Table 1. (Continued)

Country	2011 Atlantic Salmon Production (metric tons)	2011 Atlantic Salmon Value (US Dollars 000s)	2011 Exports to the United States (metric tons)	2011 Exports to the United States (US Dollars 000s)
Faroe Islands	60,473	413,999	14,935	97,208
Australia	35,198	419,850	1	-
United States	18,595	104,132	-	-
Ireland	12,196	101,987	322	3,796
Others	10,433	35,913	5,652	41,853
Total	1,721,254	9,710,782	181,225	1,520,228

Source: FAO, Fisheries and Aquaculture Department, Fisheries Statistics and Information, Global Aquaculture Production 1950-2011 Online Query, http://www.fao.org/fishery/statistics/global-aquaculture-production/query/ en.

National Marine Fisheries Service, U.S. Foreign Trade, Imports of Fishery Products, Online Query, http://www.st.nmfs.noaa.gov/commercial-fisheries/foreign-trade/raw-data/imports-exports-annual#1.

Notes:

a. U.S. imports from Chile rebounded in 2012 with an increase to 93,552 thousand metric tons. For 2009 to 2011, disease problems on Chilean fish farms decreased salmon production and imports to the United States.

AquaBounty Case: The First Genetically Engineered Food Fish

AquAdvantage Salmon

The Atlantic salmon is the first genetically engineered fish to be considered for commercial production and human consumption.[42] AquaBounty Technologies, Inc. is currently seeking regulatory approval from the FDA to sell its AquAdvantage salmon for human consumption in the United States.[43] In 2011, the company also received a grant from the U.S. Department of Agriculture's National Institute of Food and Agriculture for work on technologies to render transgenic fish sterile to reduce the risk of gene flow to wild populations.[44]

Genetic engineering technology can be used to introduce a desirable trait(s) into an organism by transferring genetic material (DNA) from another organism. The process creates recombinant DNA (rDNA)—the result of combining two or more DNA sequences that would not normally occur together in nature. Unlike the natural process of genetic recombination, rDNA is engineered by isolating and combining DNA in a laboratory. The DNA that carries the code for a desirable protein such as a hormone is then introduced to an existing organism such as an Atlantic salmon. The introduced DNA becomes part of the organism's genetic makeup and may be passed on to the organism's offspring. Some sequences of DNA are promoters which tell the organism's cells when to make certain substances. Promoters can be spliced to the desired gene that has specific instructions to make a protein such as a growth hormone. When genes are moved from one organism to another a transgenic organism is created.

GE salmon were developed by injecting rDNA composed of a promoter from another fish, an ocean pout, and a growth hormone gene from a Pacific Chinook salmon into fertilized eggs of Atlantic salmon. Subsequent selection and breeding led to the development of the AquAdvantage salmon line, which produces growth hormone throughout the year. The year-round production of growth hormone allows for continuous feeding and growth of AquAdvantagesalmon. Growth hormone production of non-GE Atlantic salmon decreases during the winter months, and Atlantic salmon stop feeding and growing during this period.

The AquAdvantage salmon also increases the efficiency of salmon production because of faster growth and better feeding efficiency than non-GE Atlantic salmon. salmon reach smolt size[45] more quickly than non-GE Atlantic salmon and grow to a market size of 1 to 3 kilograms in 16 to 18 months instead of the typical three years.[46] Although AquAdvantage salmon grow more quickly, they do not reach an overall larger size than non-GE Atlantic salmon. According to AquaBounty, analysis of AquAdvantage salmon has shown that they consume 25% less feed to achieve the same size as non-GE Atlantic salmon.[47] Feed is the most significant cost for commercial salmon aquaculture operations. Characteristics exhibited by GE salmon include accelerated growth, elevated metabolism, greater feeding motivation and efficiency, increased aggression and foraging activity, and reduced anti-predator response.[48] Similar traits have been observed for domesticated Atlantic salmon developed through selective breeding.

Faster growth confers an advantage to using GE salmon relative to non-GE salmon and, according to some, could make land-based closed aquaculture systems competitive with cage culture currently used in coastal areas. Proponents of GE salmon maintain that this is a significant development because of environmental harm caused by salmon cage culture.

AquaBounty Application

Proposed Operations

The AquAdvantage salmon would be produced and imported into the United States under specified conditions proposed by AquaBounty. AquaBounty would produce eyed eggs at a specific facility on Prince Edward Island (PEI), Canada.[49] Eggs would be shipped to Panama and reared to market size in land-based facilities. The grow-out facility would be based in the Panamanian highlands to reduce the risk of salmon escapes and interactions with wild salmon populations.[50] Salmon would be processed in Panama before being shipped to the United States for retail sale and no live fish would be imported into the United States.

AquaBouuty has stipulated that they would only produce sterile female GE Atlantic salmon by a process which manipulates salmonid reproductive biology. The production of monosex salmon is considered to be 100% effective. In addition, pressure treatment of the eggs induces triploidy (an extra set of chromosomes) which results in sterility. When done on a commercial scale, batches of eggs are on average 99.8% triploid and rates greater than 98% are expected for most inductions.[51] All-female lines of triploid fish are considered to be one of the best current methods to insure nonbreeding populations of GE fish. Therefore, the risk of an independent breeding population of GE salmon is considered to be extremely low.

Growing GE marine fish in isolated onshore tanks rather than in offshore or nearshore pens may substantially lower the risk of escape into the wild. Both facilities currently used by AquaBounty confine production to land-based freshwater areas and proposed production would be continued in this manner. The egg production facility on Prince Edward Island is currently licensed to conduct research on GE fish under Canadian regulations. The facility has incorporated redundant measures to provide for physical containment and ensure that neither brood stock nor eggs can escape. Security is also provided at the PEI facility to stop unauthorized or unintentional access.

The grow-out facility in the highlands of Panama is located at the upper portion of a watershed at 5,000 feet above sea level. The river which supplies the facility runs into several other tributaries and discharges into the Pacific Ocean. Water is diverted from the river into a basin which supplies the facility's grow-out tanks. Screens are used wherever water flows out of the facility to prevent the escape of fish while security is provided to deter human or animal intrusion. The Panama site is geographically isolated from the range of salmon species, and environmental conditions in the river's estuary and the Pacific Ocean are unfavorable for salmon survival. According to FDA's Draft Environmental Assessment, in the event that AquAdvantage Salmon escape, geographical and geophysical containment would preclude or significantly reduce the probability of survival, dispersal, and long-term establishment.

Application History

In 1993, AquaBounty first approached the FDA concerning the commercial use of GE salmon, and in 1995 they formally applied for approval. In 2009, AquaBounty provided FDA with the last required study of AquAdvantage Atlantic salmon for their New Animal Drug Application (NADA). On September 19-20, 2010, FDA's Veterinary Medicine Advisory Committee (VMAC) met to consider issues regarding the safety and effectiveness of the NADA. The public was also given the opportunity to provide written submissions and oral testimony to the committee. On December 20, 2012, FDA announced the availability for public comment of (1) the draft environmental assessment of impacts associated with NADA submitted byAquaBounty[52] and (2) FDA's preliminary finding of no significant impact (FONSI).[53] A 60-day public comment period initially ran through February 25, 2013, but was extended through April 26, 2013.[54] On November 23, 2013, Environment Canada granted AquaBounty permission to export up to 100,000 eggs a year from a hatchery in Prince Edward Island to Panama. A land-based research facility is currently operating and raising GE salmon in Panama.[55] On March 13, 2014, the FDA Commissioner, Dr. Margaret Hamburg, stated that the Aquabounty NAD application is still under consideration and that FDA will be moving forward in a deliberate science-driven way.[56]

Food Safety

The VMAC briefing pack included a section on food safety which concluded that there are no direct or indirect food consumption hazards related to AquAdvantage salmon.[57] Although the VMAC concluded that test results established similarities and equivalence between AquAdvantage salmon and non-GE Atlantic salmon, the chairman's report added that it cannot be concluded from the data submitted that AquAdvantage Salmon would be more or less allergenic than Atlantic salmon.[58] FDA has maintained that people who are allergic to Atlantic salmon will likely be allergic to AquAdvantage Salmon because it is a finfish, but not because it has been genetically engineered.[59] In the preliminary finding of no significant impact released on December 20, 2012, FDA reiterated that food from AquAdvantage salmon is as safe as food from non-GE salmon and determined that there are no significant food safety hazards or risks associated with AquAdvantage salmon.[60]

The Veterinary Medicine Advisory Committee and GE Salmon

The Center for Veterinary Medicine regulates transgenic animals intended for human consumption under the same authority it uses to regulate NADs.[61] As a first step in the review process, on September 19-21, 2010, FDA's Veterinary Medicine Advisory Committee met to consider the application and held a public hearing.[62] The FDA uses advisory committees and panels to obtain expert advice on science, technology, and policy. The VMAC is composed of members with technical expertise such as veterinary medicine, animal science, microbiology, biostatistics, and food sciences. Opponents of the AquaBounty application have been critical of the committee's composition and have argued for more experts in fisheries and ecology. The FDA charged the VMAC with reviewing issues regarding the safety and effectiveness of the new animal drug application. The VMAC Chairmen's report provided the following four basic questions and responses to issues related to AquAdvantage Salmon.[63]

1. Do the data and information demonstrate that the rDNA construct is safe to AquAdvantage salmon?
   The committee found no evidence in the data to conclude that the introduction of the construct was unsafe to the animal.
2. Do the data and information demonstrate that there is a reasonable certainty of no harm from consumption of foods derived from AquAdvantage salmon? (safety of food from AquAdvantage Salmon was considered in the context of non-GE Atlantic salmon)
   The committee deemed the studies selected to evaluate this question to be overall appropriate and a large number of test results established similarities and equivalence between AquAdvantage Salmon and Atlantic salmon.
3. Do the data indicate that AquAdvantage Salmon grow faster than their conventional counterparts?
   The committee found evidence in support of this claim.
4. Are any potential environmental impacts from AquAdvantage Salmon production adequately mitigated by AquaBounty Technologies' proposed conditions of use?
   Although the committee recognized that the risk of escape from either facility could never be zero, the multiple barriers to escape at both the Prince Edward Island and Panama facilities were extensive. Because part of the containment strategy is dependent on management standard operating procedures, the committee felt that rigorous adherence to policy would need to be maintained at both sites to sustain the barriers.

Evaluation of Potential Environmental Impacts

To assess potential environmental impacts, FDA has released a draft EAand consulted with the Fish and Wildlife Service and the National Marine Fisheries Service (NOAA Fisheries). In 2007, legislation was passed to require the FDA to consult with the National Marine Fisheries Service and to produce a report on any environmental risks associated with genetically engineered seafood products, including the impact on wild fish stocks.[64]

According to FDA, the two agencies have consulted on this matter, but this report has not been developed and no target date for its completion has been specified.

Summary of FDA Conclusions Concerning the Environmental Impacts of AquAdvantage Salmon

The Draft EA attempted to address potential hazards and harms to the U.S. environment if AquAdvantage Salmon were to escape. The EA posed the following risk related questions and provided the corresponding answers.[66]

1. What is the likelihood that AquAdvantage Salmon will escape the conditions of confinement?
   Due to multiple, redundant containment measures at the sites of egg production and grow-out, the possibility of AquAdvantage Salmon (or the broodstock used to produce these fish) escaping to the environment is extremely remote.
2. What is the likelihood that AquAdvantage Salmon will survive and disperse if they escape the conditions of confinement?
   In the unlikely event of an escape or release, environmental conditions at both the egg production and grow-out sites are sufficiently inhospitable to limit long-term survival and spread of AquAdvantage Salmon to other locations.
3. What is the likelihood that AquAdvantage Salmon will reproduce and establish if they escape the conditions of confinement?
   AquAdvantage Salmon would be produced as all-female, triploid fish. As such they would be effectively sterile. The combination of triploidy and an all-female population is expect to render AquAdvantage Salmon effectively and functionally sterile resulting in complete reproductive containment.
4. What are the likely consequences to, or effects on, the environment of the United States should AquAdvantage Salmon escape the conditions of confinement?
   The collective information on the potential for survival, dispersal, reproduction and establishment indicates that exposure pathways for AquAdvantage Salmon to reach the United States are incomplete; therefore, no effects are expected on the environment of the United States.

FDA has made a "no effect" determination under the Endangered Species Act (ESA; 16 U.S.C. 1531 et seq.) and concluded that approval of the AquAdvantage Salmon NADA will not jeopardize the continued existence of Atlantic salmon listed as threatened or endangered or result in the destruction or adverse modification of their critical habitat. FDA and NMFS engaged in technical discussions during 2010 and 2011 and based on those discussions NMFS did not object to the proposed action. However, critics have questioned whether FDA has sufficient expertise to identify and protect against all potential ecological damage that might result from the widespread use of transgenic fish.[65]

The draft environmental assessment evaluated the potential environmental impacts associated with approving the NADA for AquAdvantage Salmon. According to FDA, it has verified that AquAdvantage Salmon would be produced and grown in secure facilities. The escape and survival of GE salmon from containment into the local environments of PEI and

Panama is considered by FDA to be extremely remote. The environment around the egg producing facility and the grow-out facility are described by FDA as inhospitable. In the event that fish escape and survive, reproduction in the wild would be unlikely because the AquAdvantage salmon will be all female triploid fish which are nearly all sterile.[67] In the draft EA, FDA concluded that it "found no evidence that approval of an NADA for AquAdvantage Salmon would result in significant impacts on the environment in the United States."[68] If significant new information or challenges arise in the public comments, FDA may determine that a full environmental impact statement is required prior to approval of AquaBounty's application.

NEPA does not require an analysis of environmental effects in other countries and therefore, potential effects on the environment in Canada and Panama were not considered. These effects would be evaluated only if potential exposure pathways exist which could cause significant effects on the environment in the United States. Social and economic effects were not analyzed in the EA because the proposed action, if implemented as required, is not anticipated to significantly affect the physical environment of the United States.

In an effort to broaden the evaluation of the AquaBounty application, a coalition of environmental groups called on FDA to prepare an EIS on this action and to consult more closely with federal agencies about possible threats to endangered wild Atlantic salmon.[69] On May 25, 2011, these groups filed a formal citizen petition urging FDA to withhold approval until an EIS has been completed.[70] Some VMAC members also stated that an EIS would be needed if the company proposes additional facilities for growing salmon.

FOOD SAFETY ISSUES

A National Research Council study maintains that there is a low to moderate food safety risk from GE seafood.[71] Since genetic engineering can introduce new protein into a food product, there are concerns that this technique could introduce a previously unknown allergen into the food supply or could introduce a known allergen into a "new" food.

On February 7, 2012, three non-governmental organizations petitioned the FDA's Office of Food Additive Safety (OFAS) to review the AquaBounty application under the FFDCA food additive provisions.[72] The petitioners argued that the gene expression product (GEP) of the genetic construct creating the AquAdvantage salmon, is a food additive under FFDCA (§201(s), 21 U.S.C. §321). AquAdvantage salmon exhibit an elevated level of Insulin Growth Factor-1 (IGF-1), which they asserted is a novel food additive and constitutes a "material fact" about the GE-salmon compared to its non-GE counterpart.[73] They have requested that FDA make a finding that neither the AquAdvantage salmon nor the GEP used to create it is "generally regarded as safe" (GRAS).[74] In particular, the petition requested extensive pre-market testing, arguing that "[t]he Agency's general classification of rDNA constructs as new animal drugs does not displace or override the Agency's regulations and guidelines, and nothing precludes the Agency from also regulating GE salmon and its components as food additives."

Opponents of approval also question the validity of data supplied by AquaBounty for risk assessment.[75] They contend that FDA should have AquaBounty re-conduct its studies, or FDA should either conduct the studies themselves or ask an independent laboratory to undertake the studies. If the AquAdvantage Salmon were shown to have materially different nutritional or health characteristics from a non-GE salmon, while still being deemed safe, FDA could require that the fish be labeled.

FDA has continued to review the AquaBounty proposal solely under the NAD protocol. Because the review process of the AquAdvantage Salmon could become precedent-setting for review of other GE animals, the issue of the appropriate agency regulatory protocol—NAD or that of a food additive—is likely to remain a significant point of contention in the regulatory process.

ENVIRONMENTAL ISSUES

The potential harm that might be caused by GE organisms which escape from aquaculture facilities is of great concern to some scientists and environmental groups.[76] A National Research Council report stated that transgenic fish pose the "greatest science-based concerns associated with animal biotechnology, in large part due to the uncertainty inherent in identifying environmental problems early on and the difficulty of remediation once a problem has been identified."[77] For AquAdvantage Salmon, concerns include interbreeding and competing with wild Atlantic salmon and competition with fish both within and outside the range of Atlantic salmon.

Interbreeding with Wild Atlantic Salmon

Experiences with farmed Atlantic salmon may provide some insights regarding potential interactions of GE Atlantic salmon and wild Atlantic salmon. Farmed Atlantic salmon frequently escape from fish farms in areas both within and outside their native ranges.[78] Escaped farmed salmon have been found on spawning grounds during the period when wild Atlantic salmon spawning occurs.[79] Farmed salmon also spawn in these areas, but with lower success than wild Atlantic salmon. Successful spawning appears to have most frequently occurred between farmed females and wild males. The offspring of farmed and farmed and wild salmon (hybrids) occupy areas also inhabited by wild salmon. It is not known whether the presence of farmed and hybrid juveniles limit food and resources for wild fish and competitively displace native juveniles. The outcomes of interactions between farmed and wild juveniles are likely to vary depending on specific circumstances such as the quality and availability of habitat and associated resources.

Domestication of farmed salmon has changed their genetic composition and reduced genetic variation. These changes have occurred because limited numbers of brood fish are used for spawning farmed fish and farmers select for specific traits.[80] Much of present-day farm production of Atlantic salmon is now based on five Norwegian strains. Farmed and wild hybrids and backcrossing of hybrids in subsequent generations may change genetic variability and the frequency and type of alleles present in wild populations.[81] The extent and nature of these changes to genetic variability may affect survival (fitness) of these populations. Changes in the genetic profiles of wild populations have been shown to have occurred in several rivers in Norway and Ireland where inter-breeding of wild and farmed fish is common.[82] Large-scale experiments in Norway and Ireland show highly reduced survival and lifetime success of farmed and hybrid salmon compared to wild salmon.[83]

Given previous experiences with farmed salmon, opponents hypothesize that farmed GE salmon will eventually escape from aquaculture systems and interbreed with wild Atlantic salmon. GE salmon may exhibit different fitness-related traits such as higher feeding and growth rates. Researchers have questioned whether the flow of a gene or genes from transgenic fish such as GE salmon may confer specific advantages to hybrids relative to wild fish resulting in population-wide consequences.[84] The "Trojan gene hypothesis" speculates that populations could become extinct when a gene that confers a reproductive advantage also renders offspring less able to survive in the natural environment. However, comments to the VMAC from one of the researchers who framed the hypothesis stated that the Trojan gene effect only occurs when there is a conflict between mating success (if GE salmon were to mate more successfully) and viability fitness (offspring were less likely to survive in the wild). He concluded that the risk of harm is low because data conclusively show that in this case there is no Trojan gene effect and the transgene will be purged by natural selection.[85] However, the potential consequences of the interbreeding of farm (including GE salmon) and wild Atlantic salmon remains a long-term concern of those attempting to conserve wild Atlantic salmon populations.

Competition and Other Interactions

Critics are also concerned that AquAdvantage Salmon could become established in the wild and compete with native fish, including both Atlantic salmon and other species, for food, habitat, mates, and other resources.[86] This is a concern both within and outside the range of Atlantic salmon. Previous deliberate attempts to introduce Atlantic salmon have failed and no self-sustaining populations of anadromous Atlantic salmon have been established outside the natural range of the species. In British Columbia, Atlantic salmon which escaped from fish farms have spawned and produced wild-spawned juvenile Atlantic salmon, but it is uncertain whether they have established self-reproducing breeding populations.[87] Experimental crosses between GE Atlantic salmon and wild brown trout (*Salmo trutta*) have been shown to be viable and to confer a growth advantage to the hybrid. Researchers concluded that transgenic hybrids could detrimentally affect wild salmon populations, but that introgression of the gene into the brown trout genome through backcrossing is unlikely.[88]

If GE salmon were to escape and establish self-sustaining populations, competition for resources is another potential concern. Since GE salmon grow faster, it has been suggested that they may outcompete wild fish for habitat and food.[89] Laboratory experiments on AquAdvantage relatives indicate that they are more likely to feed in the presence of a predator than non-GE controls.[90] Another study which compared GE and non-GE salmon fry under food-limited conditions in simulated environments showed no difference in territorial dominance, growth, or survival of first feeding fry at high densities.[91] Biotechnology proponents argue that GE fish, if they escape, would be less likely to survive in the wild, especially when they are reared in protected artificial habitats and have not learned to avoid predators.

The consequences of potential competition would also depend on many factors, including the size and health of the wild population, the number and characteristics of the escaped fish, and local environmental conditions. Some argue that once transgenic fish become established, they could be difficult or impossible to eradicate such as many invasive species. This scenario would depend on reproduction of GE salmon in the wild.[92] Critics also express concerns that U.S. wild Atlantic salmon populations are at extremely low levels and especially vulnerable to ecological changes.

Future Concerns

According to the FDA, the NADA is for a specific set of conditions and any modification to these conditions would require notification of the FDA.[93] Major or moderate changes to these conditions would require a supplemental NADA which would trigger environmental analyses under NEPA. The FDA states that expansion or changes to facilities identified in the proposal would constitute major or moderate changes which would require a supplemental NADA and analysis under NEPA. If FDA approves AquaBounty's application, FDA retains the authority to withdraw its approval should significant subsequent concerns arise.

If GE salmon are approved for consumption in the United States, it is likely that additional applications will be made for additional geographic areas and for a variety of different grow-out or culture techniques. Under the current proposal, AquaBounty would produce GE salmon in a limited grow-out facility in Panama. Some have speculated that subsequent proposals would be subject to less scrutiny and question whether future uses will be as apparently benign.[94] If additional applications are approved, it is possible that future grow-out facilities would be located closer to the range of wild Atlantic salmon thereby decreasing the effectiveness of geographic containment. The NADA would not include the culture of AquAdvantage Salmon in net pens,[95] but if cage culture were used to produce GE salmon it would increase the probability of fish escaping to the wild. AquaBounty has denied that their company would approve the use of AquAdvantage Salmon in cage culture.[96]

Labeling Issues

The question of how to label the food derived from the AquAdvantage salmon is separate from the decision about whether to approve the new animal drug application. Although FDA is not required to address labeling issues prior to the food being marketed, FDA is considering these two issues simultaneously.[97] If the AquAdvantage salmon NADA is approved, FDA is to determine whether additional labeling is appropriate.

FDA has determined that the AquaBounty salmon is as safe for human consumption as non-GE salmon, in other words, that the AquaBounty salmon is "substantially equivalent" to non-GE salmon. Opponents of the AquaBounty salmon, however, have argued the need to label it as a GE product on a presumed basis of consumers' right-to-know. Proponents of the AquaBounty salmon argue that labeling a "substantially equivalent" food would imply that the GE fish was different in ways that could be seen as negative. Given that the labeling issue for GE foods remains unsettled, labeling the AquaBounty salmon, as potentially the first GE animal approved by FDA for human consumption, has become an important aspect of FDA's overall approval process. While voluntary labeling would be permissible—as long as the label was not false or misleading— proponents of the AquaBounty salmon believe such labeling could reduce retail demand for the fish. Many major grocery retail chains, such as Safeway, Trader Joe's, Target, Kroger, and Whole Foods, have already announced that they would not sell the AquaBounty fish when it becomes commercially available.[98]

States have also taken steps to regulate the transport and use of GE fish. For example, Maryland,[99] Washington,[100] Oregon,[101] Minnesota,[102] Wisconsin,[103] and California[104] have passed laws banning the release of GE fish in some or all state waters. In addition, Alaska requires GE fish to be labeled.[105] No federal law specifically addresses labeling GE fish and seafood.

FISHING INDUSTRY INTERACTIONS

The success of farm-raised Atlantic salmon has made some who work in the commercial wild salmon fishing industry, particularly those in Alaska, especially sensitive to potential impacts of GE salmon. Fishermen have concerns related to further increases in salmon aquaculture production and environmental harm to wild stocks. The salmon fishing industry in Alaska does not produce intensively farmed fish salmon in net pens and would not benefit from the use of GE salmon.[106]

Over the last three decades, the rapid growth of farmed salmon and trout production has been one of the primary drivers of world salmon prices.[107] In 2002, prices paid to Alaska fishermen were less than half of the average prices paid from 1980 to 2005.[108] Changes to salmon markets were actually much more complex during this period because the industry was transformed by a variety of factors in addition to the growth in the total supply of salmon. Salmon markets were also affected by the availability of different types of salmon products, the timing of production, and development of market standards.

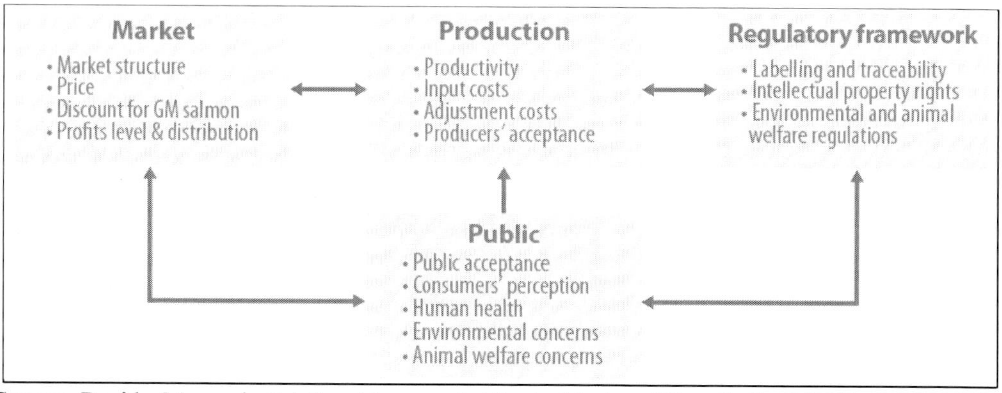

Source: Davide Menozzi, Cristina Mora, and Alberto Merigo, "Genetically Modified Salmon for Dinner? Transgenic Salmon Marketing Scenarios," *AgrBioForum*, vol. 15, no. 3 (2012).

Figure 1. Driving Forces of Salmon Production.

Some believe that the wide use of GE salmon would depress salmon prices. Potential interactions between GE salmon aquaculture, ongoing salmon aquaculture, and wild fisheries are difficult to fully assess. Part of the difficulty is related to the complex nature of interrelated drivers of aquaculture and wild salmon production. For example, public acceptance can affect prices which in turn may affect salmon production. Major categories of drivers include the regulatory framework, production, markets, and the public (**Figure 1**).[109]

Although some assume that approval of GE technology will be followed by a rapid increase in farmed salmon production, it is unlikely that this will occur.At least initially, annual production at the AquaBounty Panama facility is limited to a capacity of approximately 100 metric tons compared to current U.S. salmon consumption of approximately 180 thousand metric tons. Significant production increases would depend on approval of additional sites in salmon producing regions of the world. The regulatory framework will determine which production methods would be used and where they are allowed. If only land-based facilities are allowed, then greater aquaculture production would depend on whether the advantages of using GE salmon (higher growth rate and feed conversion efficiency) can outweigh the greater costs of land-based facilities. Currently, the economic viability of most land-based salmon grow-out facilities is questionable.[110] Significant increases in land-based production also may be constrained by the number of available sites with sufficient water quality and volume. Therefore, the magnitude and timing of greater salmon production, if it were to occur, would be very uncertain.

Some in the commercial fishing industry have stated that it has successfully educated the public to discriminate among fish from different sources, such as wild and farmed salmon. On the other hand, some believe that a publicized escape of GE fish could lead to less public acceptance of their wild product. To differentiate wild and GE salmon, commercial Alaskan fishermen also have requested the labeling of GE salmon.[111] Some industry groups are concerned that such labeling might lead consumers to believe that wild fishery products are also genetically engineered and perceived as unsafe for consumption.[112] An alternative considered by commercial interests would involve labeling of wild salmon as a product that was not produced with GE technology.

Consumer Acceptance

It is uncertain whether consumers will accept GE fish, but it appears that a broad segment of the U.S. population is opposed or at least skeptical of consuming the product.[113] GE salmon may taste the same and are expected, like other Atlantic salmon aquaculture products, to be less expensive than wild-caught fish. However, food safety issues, ethical concerns over the appropriate use of animals, and environmental concerns might affect public acceptance of GE fish as food. Ongoing campaigns by environmental and consumer groups have asked grocers, restaurants, and distributors to sign a pledge to not sell GE fish products, even if they are approved by FDA.[114]

Demand for fish products, especially those high in fish oils such as Atlantic salmon, grew quickly during the past three decades and is expected to continue growing. In the past, aquaculture productivity gains and reductions in cost have been passed on to the consumer in the form of lower prices. One study speculated that consumers would require a significant price discount to purchase the GE salmon product.[115] The discount would likely vary by country and depend on consumer demographic factors such as age.

Production increases could contribute to consumer benefits associated with lower prices and health benefits from consumption of salmon rather than less healthy protein sources. If GE salmon can be sold for a relatively low price, it could stimulate salmon consumption in low-income households.[116] Some researchers have questioned whether the potential benefits associated with the wide use of GE salmon also should be considered as part of the approval process.[117]

Congressional Actions

Some Members of Congress have raised concerns about FDA's approval process for GE salmon. On April 24, 2013, 20 Members of the House and 12 Members of the Senate sent similar letters that requested FDA Commissioner Margaret Hamburg halt the approval process.[118] In particular, the letters stated that the FDA process has not been adequate to ensure GE salmon is safe for the environment and consumers. They also expressed concern with the precedent that this ruling could set for future applications for other genetically engineered fish such as tilapia and trout. If the agency still proceeds with approval of the product, Members also urged FDA to develop labeling requirements to distinguish the product as genetically engineered.

Bills focusing on regulating or prohibiting GE salmon have been introduced in the 113th Congress. S. 246 and H.R. 1667 are similar bills which would make it unlawful to:

> (1) ship, transport, offer for sale, sell, or purchase a covered fish (GE fish), or a product containing a covered fish, in interstate or foreign commerce;
> (2) have custody, control, or possession of, with the intent to ship, transport, offer for sale, sell, or purchase a covered fish, or a product containing covered fish, in interstate commerce;
> (3) release a covered fish into a natural environment; or
> (4) have custody, control, or possession of a covered fish with the intent to release it into a natural environment.

The definition of a covered fish under H.R. 1667 would include any finfish while S. 246 would include salmon, anadromous fish, or marine fish. H.R. 1667 also would explicitly make it illegal to engage in net-pen culture of covered fish. Both bills would include exceptions for scientific research, fish collected for the purpose of supporting the act, or if NOAA finds that there would be no significant impact in accordance with NEPA. The NEPA analysis would have to include an environmental risk analysis, assessment of best or worst case probabilities of confinement failure, costs to eradicate escaped covered fish, and assessment of economic damage of escaped covered fish. Both bills also would require completion of the report on environmental risks associated with GE seafood products which was required under Section 1007 of the Food and Drug Administration Amendments Act of 2007 (21 U.S.C. 2106).

H.R. 584 and S. 248 would amend Section 403 of the FFDCA (21 U.S.C. 343) by adding a requirement to label products which contain genetically engineered fish. H.R. 1699 and S. 809 would require labeling of genetically engineered foods including fish. On May 22, 2013, S. Amdt. 965 to S. 954 was proposed and would have required that any food offered for sale have a label indicating that the food contains a genetically engineered ingredient. On May 23, 2013, the amendment was defeated by a vote of 27-71.

In the 112th Congress, several bills also were introduced to address concerns related to GE fish including House and Senate bills which would have required labeling of GE fish (S. 229, H.R. 520, and H.R. 3553), deemed GE fish as unsafe for human consumption (S. 230 and H.R. 521) and prohibited the sale of GE fish (S. 1717). On June 16, 2011, Section 744 of H.R. 2112 was passed by the House which would have prohibited the FDA from spending FY2012 funds to approve any application for GE salmon. On September 7, 2011, the Senate Committee on Appropriations reported H.R. 2112 without the prohibition on FDA spending (S. Rept. 112-73), and the provision was not included in the subsequently enacted P.L. 112-55. On May 24, 2012, S. Amdt. 2108 to S. 3187 was defeated, which would have prohibited approval of GE fish by FDA unless NOAA concurred with such approval. No further action was taken on any of these bills during the 112th Congress.

FUTURE CONSIDERATIONS

Some have asserted that the FDA's approach to GE salmon evaluation fails to consider the full scope of potential impacts or to fully assess the risk of unintended consequences. They claim that neither the VMAC nor the EA have fully considered the broader potential risks of AquAdvantage salmon approval. They have argued that FDA is not suited to undertake biological and ecological studies of this nature and they question whether the regulatory system has kept pace with advances in GE technology. They support development of a comprehensive EIS to more fully assess the broader context of social, economic, and environmental implications of GE salmon.

FDA has evaluated GE salmon as a new animal drug which is safe for human consumption. Regulation of transgenic animals as NADs, however, suggests to some observers (e.g., the Center for Food Safety, Union of Concerned Scientists) the inherent weakness of existing regulatory structures to respond adequately to the complexities that arise with animal biotechnology innovations.[119] An immediate issue is whether AquAdvantagesalmon should be subject to a more rigorous assessment as a food additive by FDA's Office of Food Additive Safety. Some might argue that the NAD assessment has already covered issues related to food safety and found that GE salmon are essentially the same as non-GE salmon. However, in addition to ensuring the safety of GE salmon further assessment could reassure consumers and retailers who are skeptical of the new technology, and more rigorous assessment would set a precedent for future products which may warrant greater scrutiny.

Some have expressed concerns that a continued delay signals that the United States might cede its leadership position in agricultural biotechnology. Rejection of theAquaBounty application would send the message that science-based regulatory oversight is subject to political intervention.[120] Other countries which are likely to commercialize this technology may be less concerned with potential environmental effects. GE technology is likely to be adopted for aquaculture in other parts of the world and according to the FDA, implications of GE salmon disapproval in the United States and adoption of the technology elsewhere are unknown.[121] Aquabounty has reported that they have spent $50 million developing the technology and in applying for approval. Some would question whether technology industries in the United States will continue to attract investors and compete with companies in other parts of the world when there is such a long and uncertain approval process.

AquAdvantage salmon is likely to be the first of many transgenic candidates for commercial aquaculture production, and some question whether this case will establish a proper or useful precedent for future assessment of this technology.[122] Some scientists assert that the FDA decision involves broader social costs and benefits of using genetic engineering and support wide-ranging interdisciplinary evaluation of GE salmon approval.[123] According to one group of researchers, assessment of long-term changes related to GE salmon will require the interrelated study of production systems, markets, consumer acceptance, and regulatory framework.[124] They also would include assessment of potential long-term changes associated with the wide use of GE salmon in major salmon-producing countries. Generally, it appears that many would prefer a comprehensive analysis rather than a piecemeal approach as new culture methods and areas are proposed by successive applications. Whether the current process affords adequate safeguards for the public while allowing for the application of new genetic technologies remains an open question.

End Notes

[1] Recombinant DNA (deoxyribonucleic acid) is formed by using laboratory methods to combine DNA from different sources to create DNA sequences that do not occur in nature.

[2] U.S. Department of Health and Human Services, Food and Drug Administration, *Guidance for Industry Regulation of Genetically Engineered Animals Containing Heritable Recombinant DNA Constructs*, Rockville, MD, January 15, 2009. Hereinafter cited as FDA Guidance 2009.

[3] I.G. Cowx et al., *Environmental Risk Assessment Criteria for Genetically Modified Fishes to be Placed in the EU Market*, European Food Safety Authority, May 27, 2010, http://www.efsa.europa.eu/en/supporting/doc/69e.pdf.

[4] For a list of some of the genetically engineered organisms under research, see Table 2-2 in National Research Council, *Animal Biotechnology: Science-Based Concerns* (Washington, DC: National Academies Press, 2002), http://books.nap.edu/books/0309084393/html/73.html#pagetop.

[5] Timothy Schwab, "Is FDA Ready to Regulate the World's First Biotech Food Animal?," *FDLI's Food and Policy Forum*, vol. 3, no. 12 (July 24, 2013). Hereinafter cited as Schwab 2013.

[6] Letter from Rebecca Wodder, President, American Rivers, Margie Alt, Executive Director, Environment America, and Vikki Spruill, President and CEO, Ocean Conservancy, et al. to Margaret Hamburg, M.D., Commissioner, FDA, November 8, 2010.

[7] Food and Water Watch, Consumers Union, and Center for Food Safety, *Petition to Deem ABT Technologies' Genetically Engineered AquAdvantage Salmon an Unsafe Food Additive*, February 7, 2012.

[8] Letter from Dale Kelly, Executive Director, Alaska Trollers Association, Brian Lynch, Executive Director, Petersburg Vessel Owners Association, and Buck Laukitis, President, North Pacific Fisheries Association, et al. to Aleta Sindelar, FDA, Center for Veterinary Medicine, April 26, 2013.

[9] Alison L. Van Eenennaam, William M. Muir, and Eric M. Hallerman, "Is Unaccountable Regulatory Delay and Political Interference Undermining the FDA and Hurting American Competitiveness?," *Food and Policy Forum*, vol. 3, no. 13 (July 24, 2013).

[10] Henry I. Miller, "With a Forked Tongue: How the Obama Administration Stymies Innovation in Food Production," *Forbes*, March 19, 2014.

[11] National Research Council (NRC), *Animal Biotechnology: Science-Based Concerns*, (Washington, DC, National Academy Press, 2002). Hereinafter cited as NRC, *Animal Biotechnology* 2002.

[12] NRC, *Animal Biotechnology*, 2002. p. 14.

[13] FDA Guidance 2009. FDA noted that much of the new guidance also will be relevant to non-heritable rDNA constructs (such as modifications intended for gene therapy); a separate guidance for non-heritable constructs might come later.

[14] Also, part of the FFDCA definition of "new animal drug" is one intended for use in animals that is not generally recognized as safe and effective for use under the conditions prescribed or recommended, and that has not been used to a material extent or for a material time.

[15] The agency states at the outset: "This guidance represents the Food and Drug Administration's (FDA's) current thinking on this topic. It does not create or confer any rights for or on any person and does not operate to bind FDA or the public. You can use an alternative approach if the approach satisfies the requirements of the applicable statutes and regulations."

[16] Sections 512 and 571 of the FFDCA and Title 21 CFR 514 of associated federal regulations.

[17] Michael P. McEvilly, "Lack of Transparency in the Premarket Approval Process for AquAdvantage Salmon," *Duke Law and Technology Review*, vol. 17, no. 2 (2013), p. 421.

[18] The FDA guidance discusses how NEPA requirements will apply to the GE animal approval process.

[19] Section 514.1(b)(14).

[20] Michael P. McEvilly, "Lack of Transparency in the Premarket Approval Process for AquAdvantage Salmon," *Duke Law and Technology Review*, vol. 11 (2013), pp. 413-433.

[21] FFDCA, Section 201(s).

[22] See, for example, Pew Foundation, *Issues in Regulation*. Beginning on p. 139, the report contains an extensive discussion on how these and several other USDA authorities might be used for oversight of animal biotechnology.

[23] APHIS in 2007 established an Animals Branch within its Biotechnology Regulatory Services "to develop a regulatory framework for the possible regulation of genetically engineered animals." Source: APHIS website, "Regulation of Genetically Engineered Animals," at http://www.aphis.usda.gov/biotechnology/news_transgenic_animals.shtml.

[24] *Federal Register*, vol. 73, no. 183, September 19, 2008, at http://edocket.access.gpo.gov/2008/pdf/E8-21917.pdf.

[25] The Nutrition Labeling and Education Act of 1990 (P.L. 101-535; 21 U.S.C. §343), which amended the Federal Food, Drug, and Cosmetic Act, requires most foods to carry nutrition labeling and requires food labels with claims about nutrient content or certain health messages to comply with specific requirements.

[26] The labeling of GE foods is required by 68 other countries.

[27] Food and Drug Administration, Background Document: Public Hearing on the Labeling of Food Made from the AquAdvantage Salmon, August 2010, http://www.fda.gov/Food/GuidanceRegulation/GuidanceDocuments RegulatoryInformation/LabelingNutrition/ucm222608.htm.

[28] For more information see CRS Report RL32809, *Agricultural Biotechnology: Background, Regulation, and Policy Issues.*

[29] "Statement of Policy: Foods Derived from New Plant Varieties." Federal Register (57 FR 22984), May 29, 1992.

[30] The basic federal guidance for regulating biotechnology products is the Coordinated Framework for Regulation of Biotechnology (51 Fed. Reg. 23302), published in 1986 by the White House Office of Science and Technology Policy. A key regulatory principle is that bioengineered products should continue to be regulated according to their characteristics and unique features, not their production method—that is, whether or not they were created through bioengineering. Under the Coordinated Framework, if a bioengineered product is "substantially equivalent" to a non-bioengineered product, no special regulations or labeling are required.

[31] U.S. Food and Drug Administration. 2001. *Guidance for Industry: Voluntary Labeling Indicating Whether Foods Have or Have Not Been Developed Using Bioengineering.* http://www.fda.gov/Food/ GuidanceCompliance RegulatoryInformation/GuidanceDocuments/FoodLabelingNutrition/ucm059098.htm.

[32] Some assert that accurate labeling would require a commodity identity preservation system extending from the farmer to the consumer. From their perspective, such a system could require extensive testing and detailed record-keeping at each step along the food supply chain. Estimates of the costs of mandatory labeling vary from a few dollars per person per year to as much as 10% of a household food bill. See G. P. Gruere and S. R. Rao, "A review of international labeling policies of genetically modified food to evaluate India's proposed rule." *AgBioForum*, vol. 10, no. 1, 2007, http://www.agbioforum.org.

[33] Significant recreational fisheries exist for these stocks as well as aboriginal rights to harvest stocks in the Pacific Northwest.

[34] Fisheries Statistics Division, *Fisheries of the United States 2012*, National Marine Fisheries Service, Office of Science and Technology, Current Fisheries Statistics No. 2012, Silver Spring, MD, September 2013, http://www.st.nmfs.noaa.gov/Assets/commercial/fus/fus12/01_front2012.pdf.

[35] Chile is a major aquaculture producer of Atlantic salmon, but there are no native anadromous salmon stocks in Chile.

[36] For example Atlantic salmon have escaped from culture sites in British Columbia and survived in the wild.

[37] Ken Coons, "Global Salmon Initiative does outreach at Seafood Expo," *Seafood.com*, March 18, 2014.

[38] Gunnar Knapp, Cathy A. Roheim, and James L. Anderson, *The Great Salmon Run: Competition Between Wild and Farmed Salmon*, TRAFFIC North America, Washington DC, 2007.

[39] United Nations Food and Agriculture Organization, Fishery Statistical Collections, Global Aquaculture Production, Accessed October 15, 2013, http://www.fao.org/fishery/statistics/global-aquaculture-production/en. Hereinafter cited as FAO Global Aquaculture Production 2013. In this case, value is the revenue obtained by the aquaculturalist from the first buyer of the product.

[40] FAO Global Aquaculture Production 2013.

[41] U.S. Department of Commerce, National Oceanic and Atmospheric Administration and National Marine Fisheries Service, *Fisheries of the United States 2012*, Silver Spring, MD, September 2012, http://www.st.nmfs.noaa.gov/commercial-fisheries/fus/fus12/index.

[42] Glofish™ is a genetically altered version of the popular aquaria zebrafish (*Danio rerio*). The zebrafish was made fluorescent after the insertion of a sea anemone gene into the zebrafish egg.[42] This fish is currently legal to be sold in all states except California. Since Glofish™ are not meant for human consumption, FDA determined that the Glofish™ was not under its jurisdiction.[42]

[43] AquaBounty Technologies, Inc., http://www.AquaBounty.com/.

[44] See http://www.nifa.usda.gov/newsroom/news/2011news/brag_awards.html.

[45] Atlantic salmon go through several stages including egg, alevin, fry, parr and smolt. Salmon make the transition from fresh to salt water environments during the smolt stage. The age of this transition may vary widely depending on Atlantic salmon stock.

[46] According to the Development Fund, the Norwegian aquaculture industry has developed strains of salmon through selective breeding that grow as fast or faster than AquAdvantage salmon. See http://www.utviklingsfondet.no/files/uf/ documents/GMO-Salmon_Fast_Growing_Hype_web.pdf.

[47] "As Final U.S. decision Nears, A Lively Debate on GM Salmon," *Environment 360*, April 18, 2013, p. http://e360.yale.edu/feature/as_final_us_decision_nears_a_lively_debate_on_gm_salmon/2641/. Hereinafter cited as Environment 360, 2013.

[48] U.S. Food and Drug Administration, *AquaAdvantage Salmon*, Draft Environmental Assessment, May 4, 2012, http://www.fda.gov/downloads/AnimalVeterinary/DevelopmentApprovalProcess/GeneticEngineering/GeneticallyEngineeredAnimals/UCM333102.pdf. Hereinafter cited as FDA Draft Environmental Assessment 2012.

[49] Questions have been raised concerning AquaBounty's approval by Environment Canada for producing and transporting the GE salmon eggs. See http://lists.cban.ca/pipermail/cban-e-news/2011-October/000346.html.

[50] There are no native populations of salmon in rivers or adjacent marine areas of Panama because these areas are not suitable for wild Atlantic salmon.

[51] Food and Drug Administration Center for Veterinary Medicine, *Briefing Packet AquaAdvantage Salmon*, September 20, 2010, p. 57.

[52] FDA Draft Environmental Assessment 2012.

[53] U.S. Food and Drug Administration, *Preliminary Finding of No Significant Impact AquAdvantage Salmon*, May 4, 2012, http://www.fda.gov/downloads/AnimalVeterinary/DevelopmentApprovalProcess/GeneticEngineering/GeneticallyEngineeredAnimals/UCM333105.pdf.

[54] 78 *Federal Register* 10620-10621 (February 14, 2013).

[55] It has been reported that if the FDA approves imports of GE salmon, AquaBounty will request Panama's permission to convert the research facility into a production facility.

[56] U.S. Congress, Senate Committee on Health, Education, Labor, and Pensions, *Protecting the Public Health: Examining FDA's Initiatives and Priorities*, 113th Cong., 2nd sess., March 13, 2014.

[57] Food and Drug Administration, *Briefing Packet AquaAdvantage Salmon*, FDA Center for Veterinary Medicine, September 20, 2010, p. 61, http://www.fda.gov/downloads/AdvisoryCommittees/CommitteesMeetingMaterials/VeterinaryMedicineAdvisoryCommittee/UCM224762.pdf.

[58] Chairmen's Report, 2010.

[59] Food and Drug Administration, *Key Facts Commonly Misunderstood*, http://www.fda.gov/AdvisoryCommittees/CommitteesMeetingMaterials/VeterinaryMedicineAdvisoryCommittee/ucm226223.htm.

[60] Food and Drug Administration, *Preliminary Finding of No Significant Impact*, AquAdvantage Salmon, May 4, 2012, http://www.fda.gov/downloads/AnimalVeterinary/DevelopmentApprovalProcess/GeneticEngineering/GeneticallyEngineeredAnimals/UCM333105.pdf.

[61] The FDA Center for Veterinary Medicine's "Questions and Answers about Transgenic Fish," http://www.fda.gov/AnimalVeterinary/NewsEvents/FDAVeterinarianNewsletter/ucm133255.htm.

[62] Background documents for this public hearing are available at http://www.fda.gov/AdvisoryCommittees/CommitteesMeetingMaterials/VeterinaryMedicineAdvisoryCommittee/ucm201810.htm. Also see 75 *Federal Register* 52602-52605, August 26, 2010.

[63] David Senior, Interim Chair, VMAC, *Chairman's Report*, Food and Drug Administration, VMAC Meeting, September 20, 2010, http://www.fda.gov/downloads/AdvisoryCommittees/CommitteesMeetingMaterials/VeterinaryMedicineAdvisoryCommittee/UCM230467.pdf.

[64] 21 U.S.C. §2106.

[65] See the Center for Food Safety's "Genetically Engineered Fish," http://www.centerforfoodsafety.org/geneticall3.cfm.

[66] Food and Drug Administration, *AquAdvantage Salmon*, Draft Environmental Assessment, May 4, 2012, http://www.fda.gov/downloads/AnimalVeterinary/DevelopmentApprovalProcess/GeneticEngineering/GeneticallyEngineeredAnimals/UCM333102.pdf. Hereinafter cited as Draft Environmental Assessment, 2012.

[67] Triploid salmon have an extra set of chromosomes and triploidy renders them sterile. On average, treatment is expected to result in 99.8% triploid fish, and according to FDA with rates of over 98% for most inductions.

[68] Draft Environmental Assessment, 2012.

[69] Letter from Rebecca Wodder, President, American Rivers, et al. to Margaret Hamburg, Commissioner, U.S. Food and Drug Administration, November 8, 2010, http://www.eenews.net/assets/2010/11/08/document_pm_01.pdf.

[70] Ocean Conservancy et al., *Citizen Petition Regarding AquaBounty Technologies' Application for Approval of Genetically Engineered Salmon*, Citizen Petition Before the U.S. Food and Drug Administration, May 25, 2011, http://earthjustice.org/sites/default/files/FinalGESalmonCitizenPetition.pdf.

[71] National Research Council, *Safety of Genetically Engineered Foods: Approaches to Assessing Unintended Health Effects,* (Washington, DC: National Academies Press, 2004), http://books.nap.edu/openbook.php?record_id=10977& page=R1.

[72] Food and Water Watch, Consumers Union, and The Center for Food Safety, *Petition to Deem ABT Technologies' Genetically Engineered AquaAdvantage Salmon An Unsafe Food Additive*, February 7, 2012.

[73] IGF-1 is a hormone that helps accelerate the growth of the bioengineered salmon. According to the petition, IGF-1 has been shown in some studies to be linked to breast, colon, prostate, and lung cancers.

[74] The petition also argues that the AquaBounty salmon raises safety concerns because the salmon exhibits higher levels of the growth hormone IGF-1 that could have increased allergen potential.

[75] Tim Schwab, "Is FDA Ready to Regulate the World's First Biotech Food Animal?" *Food and Drug Policy Forum*, vol. 3 (July 24, 2013).

[76] See also Matthew Morgan, "The AquAdvantage Salmon: Who Owns Escaped Genetically Modified Animals?" *Ocean and Coastal Law Journal*, v. 17, no. 1 (2011): 127-161.

[77] National Research Council, *Animal Biotechnology: Science-Based Concerns* (Washington, DC: National Academies Press, 2004), http://books.nap.edu/openbook.php?record_id=10418&page=R1.

[78] Eva B. Thorstad, Ian A. Fleming, and Philip McGinnity, *Incidence and impacts of escaped farmed Atlantic salmon Salmo salar in nature*, Norwegian Institute for Nature Research, Norway, 2008. Hereinafter cited as Thorstad et al. 2008. For example, Atlantic salmon have escaped from areas within their ranges such as coastal farms in the North Atlantic (Canada, Norway, and Great Britain) and from areas outside their range in the Pacific (Chile and British Columbia).

[79] There are no recorded cases of hybridization in nature between Atlantic and Pacific salmonid species.

[80] Oystein Skaala, Vidar Wennevik, and Kevin A. Glover, Evidence of temporal genetic change in wild Atlantic salmon, *Salmo salar*, populations affected by farm escapees, "ICES Journal of Marine Sciences," vol. 63 (2006), pp. 1224-1233. Fitness can be generally described as the ability to survive and reproduce.

[81] An allele is one of two or more versions of a gene occupying a specific spot on a chromosome that control a specific trait.

[82] Thorstad et al. 2008.

[83] Thorstad et al. 2008.

[84] Kelly M. Pennington, Anne R. Kapuscinski, and Michale Morton, et al., "Full Life-cycle assessment of gene flow consistent with fitness differences in transgenic and wild-type Japanese medaka fish (*Oryzias latipes*)," *Environmental Biodiversity Research*, vol. 9 (2010), pp. 41-57.

[85] Draft Environmental Assessment 2012, p. 91.

[86] Atlantic salmon and Pacific salmon are unable to spawn because they are different species.

[87] Thorstad et al. 2008, p. 67.

[88] Krista B. Oke, Peter A. H. Westley, and Darek T. R. Moreau, et al., "Hybridization between genetically modified Atlantic salmon and wild brown trout reveals novel ecological interaction," *Proceedings of the Royal Society: Biological Sciences*, vol. 280, no. 1763 (August 7, 2013).

[89] Draft Environmental Assessment 2012, p. 91.

[90] Draft Environmental Assessment 2012, p. 92.

[91] Draft Environmental Assessment 2012 p. 92.

[92] Wild reproduction of GE fish is unlikely if only sterile triploid females are produced.

[93] FDA Environmental Assessment 2012, p. 2.

[94] Northwest Environmental Defense Center, *The Newsletter of the Northwest Environmental Defense Center*, Critique of Genetically Engineered Salmon, 2013, http://law.lclark.edu/live/files/14124-articlecritique-of-genetically-engineered-salmon-.

[95] Grow-out in net pens is less costly than culture of salmon in land-based facilities.

[96] Environment 360, 2013.

[97] FDA, "Background Document: The VMAC Meeting on Science-Based Issues Associated with AquAdvantage Salmon," http://www.fda.gov/AdvisoryCommittees/CommitteesMeetingMaterials/VeterinaryMedicineAdvisoryCommittee/ucm222712.htm.

[98] "Kroger, Safeway Will Nix GMO Salmon Regardless of FDA Decision," *Food Safety News*, March 4, 2014.

[99] Maryland Natural Resources Code Ann.§4-11A-02 (2003).

[100] Washington Administrative Code 220-76-100 (2003).

[101] Oregon Administrative Rules 635-007-0595 (2004).

[102] Minn. Stat. §§17.4982, 18B.285, §18B.01, 18F.01-.13 (2003).

[103] Wis. Stat. §146.60 (2002).

[104] California Fish & Game Code §15007 (2003) and Dept. of Fish and Game §671.1.

[105] AK Food & Drug Code §17.20.040 (2005).

[106] Alaska salmon fisheries benefit from releases of salmon from hatcheries which grow-out in the ocean and enhance wild stocks.

[107] Gunnar Knapp, Cathy A. Roheim, and James L. Anderson, *The Great Salmon Run: Competition Between Wild and Farmed Salmon*, Traffic North America, Washington DC, January 2007. Hereinafter cited as the Great Salmon Run 2007.

[108] Great Salmon Run 2007. p. 213.

[109] Davide Menozzi, Cristina Mora, and Alberto Merigo, "Genetically Modified Salmon for Dinner? Transgenic Salmon Marketing Scenarios," *AgrBioForum*, vol. 15, no. 3 (2012).

[110] Production from salmon land-based systems is insignificant, but the David Suzuki Foundation has advocated for land-based systems and maintains that they are viable.

[111] Erika Bolstad, "Alaskans say GM salmon threatens wild market as public FDA objections top 33,000," *Anchorage Daily News*, March 6, 2013.

[112] See CRS Report RL32809, *Agricultural Biotechnology: Background, Regulation, and Policy Issues*, by Tadlock Cowan.

[113] Shane O'Halloran, "Consumers not interested in genetically modified salmon," *Food Engineering*, April 23, 2013, pp. http://www.foodengineeringmag.com/articles/90611-consumers-not-interested-in-genetically-modified-salmon.

[114] See the Center for Food Safety's website and featured actions, "Tell Costco to Reject GE Salmon," http://www.centerforfoodsafety.org/geneticall3.cfm.

[115] Davide Menozzi, Cristina Mora, and Alberto Merigo, "Genetically Modified Salmon for Dinner? Transgenic Salmon Marketing Scenarios," *AgrBioForum*, vol. 15, no. 3 (2012).

[116] Martin D. Smith et al., "Genetically Modified Salmon and Full Impact Assessment," *Science*, v. 330 (November 19, 2010): 1052-1053. Hereinafter cited as Smith et al. 2010.

[117] Smith et al. 2010.

[118] Letter from Don Young et al. to Dr. Margaret Hamburg, Commissioner, FDA, April 24, 2013, and Letter from Senator Mark Begich et al. to Dr. Margaret Hamburg, April 24, 2013.

[119] See Center for Food Safety website, at http://www.centerforfoodsafety.org/geneticall7.cfm; and the Union of Concerned Scientists website, at http://www.ucsusa.org/food_and_agriculture/.

[120] Alison L. Van Eenennaam, William M. Muir, and Eric M. Hallerman, "Is Unaccountable Regulatory Delay and Political Interference Undermining the FDA and Hurting American Competitiveness?," *Food and Policy Forum*, vol. 3, no. 13 (July 24, 2013).

[121] Draft Environmental Assessment 2012.

[122] Smith et al. 2010.

[123] Smith et al. 2010.

[124] David Menozzi, Christena Mora, and Alberto Merigo, "Genetically Modified Salmon for Dinner? Transgenetic Salmon Marketing Scenarios," *AgBioForum*, vol. 15, no. 3 (2012).

In: Genetically Engineered Salmon: Background and Issues
Editor: Diedra P. Tharp
ISBN: 978-1-61122-558-7
© 2014 Nova Science Publishers, Inc.

Chapter 2

BRIEFING PACKET AQUADVANTAGE SALMON*

Veterinary Medicine Advisory Committee

LIST OF ACRONYMS AND CONVENTIONS EMPLOYED

~	approximately
ABRAC	Agricultural Biotechnology Research Advisory Committee
ABT	Aqua Bounty Technologies, Inc.
AFP	antifreeze protein
amp^r	ampicillin resistance
BOD	biochemical oxygen demand
bla	β-lactamase
bp	base-pair
CEQ	Council on Environmental Quality
CFR	Code of Federal Regulations
COD	chemical oxygen demand
CVM	Center for Veterinary Medicine
DHHS	US Department of Health and Human Services
cDNA	complementary deoxyribonucleic acid
DNA	deoxyribonucleic acid
dsDNA	double stranded deoxyribonucleic acid
DO	dissolved oxygen (content) EA environmental assessment
EIS	environmental impact statement
EC	Environment Canada
EO-1α	the integrated form of the AquAdvantage transgene
EPA	US Environmental Protection Agency
ERA	early-rearing area

* This is an edited, reformatted and augmented version of a document issued by the FDA Center for Veterinary Medicine, September 20, 2010.

ESA	Ecological Society of America
EU	European Union
FAO	Food and Agricultural Organization (of the United Nations)
FCR	feed conversion ratio
FDA	US Food and Drug Administration
FFDCA	Federal Food, Drug, and Cosmetic Act
FL	fork length; length of a fish from nose to tail-fork
FONSI	finding of no significant impact
FWS	US Fish and Wildlife Service
GLP	good laboratory practices
GOA	grow-out area
GE	genetically engineered
GH	growth hormone
mRNA	messenger RNA
op	ocean pout promoter regulatory region
PIT	passive integrated transponder
rDNA	recombinant deoxyribonucleic acid
USC	United States Code
USDA	US Department of Agriculture
UTR	untranslated region

TECHNICAL TERMS

AquAdvantage construct	The recombinant DNA construct used to generate AquAdvantage Salmon.
Arctic Char	A related salmonid to Atlantic salmon, used by ABT for biological containment purposes when producing AquAdvantage Salmon.
Biological Containment	The practice of constructing a genotype or environment that restricts the available locations a GE salmon can thrive.
°C-day [min]	Compound unit of "time" (°C x days [min]) for relative determination of growth rate that accounts for effect of water temperature.
Diploid	Having two complete sets of chromosomes per somatic cell.
EO-1	The mosaic, female founder of the AquAdvantage Salmon line created by microinjection of the opAFP-GHc2 transgene into a fertilized egg.
Egg	Unfertilized haploid sex cells of female salmon
EO-1α	Functional, stably integrated form of opAFP-GHc2 in the AquAdvantage Salmon genome.
Expression	Scientific terminology for the process of cellular protein production.
Flow cytometry	Method used to confirm ploidy by determination of DNA content in a dye-labeled cell population via relative fluorescence intensity.
Gamete	Haploid reproductive cells produced in sexually mature organisms.

Genome	The complete linear genetic sequence and derived information for an organism. Often refers to populations of organisms or a consensus sequence representing many organisms.
Genotype	All specific hereditary information pertaining to a specific cell, tissue, or organism. Refers to individual genes or individuals within a population.
Haploid	Having one half the normal amount of chromosomes per cell. Found exclusively in the gametes of salmonids.
Hemizygous	Having one copy of a given (trans)gene.
Homozygous	Having matching, homologous copies of a particular (trans)gene on each of two paired chromosomes in a diploid genotype.
Milt	Haploid reproductive cells of male salmon (aka 'sperm')
Molecular Cloning	A set of techniques and biotechnological tools employed to modify, assemble, and disassemble genetic sequences from one or more species with the explicit purpose of eliciting a specific biological response in a host cell or organism. Encountered here as the rDNA construct.
Neomale	A genetically female fish converted to a phenotypic male by hormone treatment.
opAFP-GHc2	AquAdvantage recombinant DNA construct comprising regulatory sequences from an ocean pout AFP gene & GH-coding sequences from chinook salmon.
PCR	Polymerase chain reaction; A common DNA amplification method used to confirm genotype by primer-extension and identification of select sequences unique to EO-1α.
Phenotype	The expression of an organism's genotype. An organism's actual observed properties, such as morphology, development, or behavior, which derive from its genotype.
PIT	Passive integrated transponder; implantable radio-beacon for fish identification.
Plasmid	A class of episomal bacterial DNAs employed in the molecular cloning of small to mid-size DNA fragments.
Ploidy	The number of complete sets of chromosomes contained within each cell of each salmon.
Promoter	A regulatory region of DNA that usually abuts a gene's protein coding sequence. Promoters regulate the state ('On/Off') and expression level ('How much protein is produced') of a gene.
Protein-coding sequence	The DNA sequence of a gene that is transcribed into mRNA and subsequently translated into protein.
Regulatory sequence	Non-protein coding DNA sequence of a gene controlling its expression.
Salmonid	The taxonomic family for andramadous, predatory ray-finned fish, including char, trout, and salmon.
Smolt	A freshwater juvenile Atlantic salmon that has undergone the physiological changes necessary to be able to survive in salt water.

Somatic	Having to do with all non-gametic tissues of an organism
SW	Sea winter: Number of winters spent at sea (e.g., 1SW, 2SW).
Transgene	Synthetic gene comprising regulatory & coding sequences constructed *in vitro* and incorporated into the genome of an organism with the intended purpose of modifying its phenotype.
Triploid	Having three complete sets of chromosomes per cell.

I. Background

A. Introduction

Genetically engineered (GE) animals have been produced since the late 1970s and early 1980s when Jaenisch, Brinster, and Palmiter (Jaenisch and Mintz, 1974; Palmiter et al., 1982a; Palmiter et al., 1982b) all reported on the development of GE mice. Not long thereafter, it was demonstrated that rabbits and pigs could also be genetically engineered (Hammer et al., 1985). Now, more than two decades later, many different species, including those traditionally consumed as food, have been engineered with various recombinant DNA (rDNA) constructs.

GE animals currently being developed can be divided into several broad classes based on the intended purpose of the genetic modification: (1) to enhance food quality or agronomic traits (e.g., pigs with less environmentally deleterious wastes, faster growing fish); (2) to improve animal health (e.g., disease resistance); (3) to produce products intended for human therapeutic use (e.g., pharmaceutical products or tissues for transplantation; these GE animals are sometimes referred to as "biopharm" animals); (4) to enrich or enhance the animals' interactions with humans (e.g., hypo-allergenic pets); (5) to develop animal models for human diseases (Tamashiro et al., 2002) (e.g., pigs as models for cardiovascular or inflammatory diseases); and (6) to produce industrial or consumer products (e.g., fibers for multiple uses).

In January of 2009, following a formal notice and comment period, FDA[1] issued Guidance for Industry 187: Regulation of Genetically Engineered Animals Containing Heritable Recombinant DNA Constructs, http://www.fda.gov/downloads/AnimalVeterinary/GuidanceComplianceEnforcement/Guidan ceforIndustry/UCM113903.pdf. For the purpose of the guidance, FDA defined "genetically engineered (GE) animals" as those animals modified by rDNA techniques, including all progeny that contain the modification. The term GE animal can refer both to animals with a heritable rDNA construct and to an animal with a non-heritable rDNA construct (e.g., a construct intended as therapy for a disease in that animal).

FDA regulates GE animals under the new animal drug provisions of the Federal Food Drug and Cosmetic Act (FFDCA or the Act), 21 USC 321 et seq., and the National Environmental Policy Act (NEPA). Section 201(g) of FFDCA defines drugs as "articles (other than food) intended to affect the structure or any function of the body of man or other animals." The rDNA construct in the resulting GE animal is thus a regulated article that meets the drug definition; the GE animal itself is not a drug. As a short-hand, the agency sometime refers to regulating the GE animal. All GE animals are captured under these provisions, regardless of their intended use.

B. Risk-Based Approach to Assessing GE Animals

FDA has developed a new hierarchical risk-based approach to assess GE animals and their edible products. It does not rely on a single "critical" study, but rather on the cumulative weight of the evidence provided by all of the steps in the review. It is risk-based because it examines both the potential hazards (that is, components that may cause an adverse outcome) identified at each step along the hierarchical pathway and likelihood of harm among the receptor populations (that is, those individuals or populations exposed to the GE animal(s) or their products).

Consistent with other FDA reviews of the products of biotechnology, this approach is, in general, "event-based." An event can be defined as the result of an insertion(s) of a recombinant DNA construct that occurs as the result of a specific introduction of the DNA to a target cell or organism. Animals derived from different events, even if they are based on the previously approved construct(s), would require separate evaluations.

Weight-of-evidence evaluation

In our weight-of-evidence evaluation of GE animals, we draw on data from a number of sources. These include the following, listed in rank order (from highest to lowest) of importance in the overall weight-of-evidence evaluation: (1) controlled studies conducted on the specific animals being considered for approval; (2) other non-controlled studies on these same animals; (3) historical records and data for these animals; and (4) studies reported in the scientific literature investigating these same animals or their relatives. Each source, in turn, is given appropriate deference with respect to its relevance to the risk or hazard identification question under consideration. Irrespective of the source or order of deference given to a given dataset, all of the data and information is evaluated in the context of basic scientific principles and external validity.

Step 1: Product Definition

The hierarchical process is based on a product definition, which in turn drives subsequent data generation and review. Product definitions ultimately characterize the GE animal intended to enter commerce, and should include the following: the ploidy and zygosity of the GE animal; a description of the animal, including the common name, genus and species; the name and number of copies of the rDNA construct; the location of the insert; the name of the GE animal line; and the claim being made for the animal. CVM recommends that sponsors identify the GE animal's genomic DNA sequences flanking the integration site(s) of the inserted rDNA to protect their intellectual property. The construct may also be given a proprietary name for similar protection.

Step 2: Molecular Characterization of the Construct

CVM recommends that sponsors provide fundamental information for identifying and characterizing the rDNA construct intended to be introduced into the GE animal intended for marketing. In general, information should be provided to describe the purpose of the modification; source(s) of the introduced DNA; details of how the rDNA construct was assembled; the intended function(s) of the introduced DNA; the sequence of the introduced DNA; and its purity prior to introduction into the initial animal or cell to be used as a nuclear donor to produce an animal via nuclear transfer.

Table 1. Weight-of-Evidence Evaluations for GE Animals

General Considerations for evaluating all evidence			
• Basic principles of science (biological plausibility) • External validity			
Order of Deference	**Description**	**Considerations**	**Example**
1	• Controlled Studies • Final Structure of rDNA Construct • Same Animal Lineage • Internal validity	• Study quality • Generation of Animals • Relevance of endpoint to risk question	• Large double blind • Use of "Good Study Practices" • Full data set • Agreed study design

2	• Non-controlled Studies
	  • Same lineage animal
	  • Same rDNA construct
		• Study size/duration
		• Study quality
		• Generation
		• Husbandry
			• Pilot study
			• Very small
			• Different endpoint
			• Summary data only
			• No study design or design not followed

3	• Historical Summary
	  • Same lineage animal
	  • Same rDNA construct
	• "Epidemiology" Study
		• Generation
		• Husbandry
			• Not "formal study"
			  • Lab/cage records
			  • Incomplete records

4	• Studies on Similar or Related Animals or Constructs
	• Different rDNA Copy Number/ Event
		• Degree of similarity
		  • Related article
		  • Related animal
			• Different species, breeds
			• Different gene confers same phenotype
			• Other regulatory elements in rDNAconstruct

Step 3: Molecular Characterization of the GE Animal

In this step, FDA evaluates the data and information supplied on the event that identifies and characterizes the subsequent GE animal, the production of the GE animal(s) intended to enter commerce, and the potential hazards that may be introduced into the animal as part of its production. Key data and information include the method by which the rDNA construct was introduced into the initial GE animal, whether the resulting animal was chimeric, and the nature of the breeding strategy used to produce the lineage progenitor.

The lineage progenitor is defined as the animal from which the animals intended to be commercialized are derived; it contains the final stabilized version of the initial event. To characterize this key animal, sponsors should provide information on the genomic location(s) of the rDNA construct's insertion site(s); number of copies of the rDNA construct at each insertion site; whether the insertion occurs in an active transcriptional region; and whether analysis of flanking sequences can help determine whether harm is likely to result from the interruption of a coding or regulatory region (insertional mutagenesis).

Step 4: Phenotypic Characterization of the GE Animal

In this and the following steps, the agency seeks to determine whether any production of the GE animal poses any public health risks (risks to human health, risks to animal health, or risks to the environment). It does so by evaluating the expression of the introduced trait and its effect(s) on the resulting GE animal. First evaluated are the data that characterize whether the rDNA construct or its expression product(s) cause any direct toxicity – that is, whether there are any adverse effects attributable to the intrinsic toxicity of the construct or its expression product(s). Indirect effects also are evaluated (indirect effects are those that may be caused by the perturbations of physiological systems by the construct or its expression product(s) (e.g., the expression product may change the expression level of another protein). In general, CVM recommends that sponsors compile and submit data and information addressing the health of the GE animals, including veterinary and treatment records, growth rates, reproductive function, and behavior. In addition, CVM recommends that data on the physiological status of the GE animals, including clinical chemistry, hematology, histopathology, and post-mortem results, be submitted for evaluation.

Step 5: Durability: Genotypic and Phenotypic Plan

This step is intended to provide information to ensure that the specific event defining the GE animal being evaluated is durable — that is, that there is a reasonable expectation that the gene construct is stably inherited and that the phenotype is consistent and predictable. FDA's specific intention for this step is for the sponsor to provide a plan to ensure that the GE animals for which data are submitted and evaluated for approval are equivalent to those intended for distribution in commerce over the commercial lifetime of the GE animal (or its products). Particular attention should be paid to the identification of GE animals derived immediately from the lineage progenitor, and the preservation of genetic material that could be used to regenerate the genetic line of the lineage progenitor, if necessary. As part of the plan, CVM recommends that sponsors maintain accurate and comprehensive records of their breeding strategy, as well as the actual breeding.

For genotypic stability, CVM recommends that sponsors use the results of studies demonstrating that the inserted transgene is consistently inherited. To demonstrate phenotypic durability, CVM recommends that sponsors submit data on the consistency of the expressed trait (based on the claim being made) over multiple generations. CVM recommends that sponsors gather data on inheritance and expression from at least two generations, preferably more, and recommends that at least two of the sampling points be from non-contiguous generations (e.g., F_2 and F_4).

The Durability Plan is inextricably linked to post-approval reporting requirements. These generally include information on the quantity of the regulated article (interpreted as the quantity of GE animals produced), any adverse events that have been reported, and any changes that may be made to the product (the GE animal). It is developed if a positive decision should be made on approving an application, and will take into account the nature and structure of the durability plan.

Step 6: Food/Feed/Environmental Safety

a. Food/Feed Safety

The food and feed safety step of the hierarchical review process addresses the issue of whether food or feed from the GE animal poses any risk to humans or animals consuming edible products from GE animals compared with the appropriate non-transgenic comparators.

The risk questions involved can be divided into two overall categories. The first asks whether there is any direct toxicity, including allergenicity, via food or feed consumption associated with the expression product of the construct or components of the construct. The second category of questions addresses potential indirect toxicity associated with both the transgene and its expressed product (e.g., will expression of the transgene affect physiological processes in the resulting animal such that unintended food/feed consumption hazards are created, or existing food/feed consumption risks are increased). Potential adverse outcomes via the food/feed exposure pathway can be identified by (1) determining whether there are any biologically relevant changes to the physiology of the animal (assessed partly in *Step 3: Phenotypic Characterization of the GE Animal*), and (2) whether reasons for toxicological concern are suggested by any biologically relevant changes in the composition of edible products from the GE animal compared with those from the appropriate non-transgenic comparator.

b. Environmental Safety

Because of the requirements set forth in the National Environmental Protection Act (NEPA) and FDA environmental impact regulations in 21 CFR 25, the agency typically must prepare an environmental assessment (EA) for each NADA approval action. The EA generally focuses on potential impacts related to the use and disposal of the GE animal. In general, the EA should describe and discuss the following: (1) the genotype, phenotype and general biology of the GE animal; (2) potential sources and pathways of escape (or release) and spread of the GE animal; (3) the types and extent of physical and biological confinement, if any that will be implemented; and (4) the potentially accessible ecosystems and their characteristics. CVM recommends that the sponsor contact CVM before proceeding with preparation of the EA in order to insure that it is appropriately focused. In the event that the EA results in a finding that a significant environmental impact may result, an Environmental Impact Statement may need to be prepared.

Step 7: Claim Validation

The previous steps of the hierarchical review approach primarily address identity and safety issues. In the last step of pre-market review, the "effectiveness" portion of the proposed claim for the GE animal is validated. In order to demonstrate effectiveness, sponsors must present substantial evidence—that is, one or more adequate and well controlled investigations (21 U.S.C. 360b(d)(3)) to validate the claim that is being made. Because the product definition contains the eventual claim, CVM recommends that sponsors contact the Center early in the development of the GE animal to reach agreement on (1) what would constitute a suitable claim, and (2) the nature and conduct of studies that would validate that claim.

C. Team-Based Review

The assessment of an application for approval by FDA is performed by an interdisciplinary team of agency subject-matter experts drawn from across the Center. In general, these teams include molecular biologists, animal scientists, veterinarians, toxicologists, chemists, statisticians, risk assessors, and other specialists, as required by the technical nature of each component of the hierarchical review process. In general, this interdisciplinary team is assembled as sponsors begin their interactions with the agency. Depending on the nature of the submission, in-depth reviewers (at least two) are assigned to each submission, each of whom prepares an initial individual evaluation of the data and information. For example, the components that address the characterization of the construct generally have molecular biologists acting as in-depth reviewers, while the phenotypic characterizations could have veterinarians, animal scientists, and statisticians as in-depth reviewers. The evaluations performed by the in-depth reviewers are presented to the full team, which has had the submission available for review; the larger group acts as a peer-review panel for the in-depth reviewers' evaluations. Following discussion, the in-depth reviewers prepare a written review, which is again subjected to peer review by the entire group. Once concurrence is reached, the entire team signs off on the review. Each step in the hierarchical review process is carried out in the same manner.

D. Transparency and Public Participation

The agency is interested in increasing the transparency of its decision-making process. To that end, after FDA has completed its review of the data and information to demonstrate safety and effectiveness, the agency intends to hold a public veterinary medicine advisory committee meeting to present its findings and receive input from the committee, as well as comments from the public. Once the agency has considered both the committee recommendation and the public comments, it can issue a statement regarding approval.

E. Summary

FDA regulates the products of the two newest forms of animal biotechnology in different ways. Cloning is considered to fall on the continuum of assisted reproductive technologies. Sufficient data were available for the agency to determine that food from cattle, swine, and goat clones is as safe to eat as food from their sexually reproduced counterparts (Walker et al., 2007; Watanabe and Nagai, 2008, 2009). The sexually reproduced offspring of clones are the same as any other sexually reproduced animals, and food from the sexually reproduced offspring of clones is the same as food from any other sexually reproduced animals. At this time, in order to ensure a smooth transition to the market, the USDA has requested that producers of clones continue to keep food from clones out of the general food supply. Food from the sexually reproduced offspring of clones has been entering the food supply freely.

Genetically engineered animals, on the other hand, are regulated under the new animal drug provisions of the FFDCA, and as such must receive formal approval before they may be introduced into commerce. The agency has issued a Guidance for Industry clarifying its statutory authority to regulate GE animals and a set of recommendations for how data and information may be submitted to the agency for review of applications for approval. The agency stresses that, due to the case by case nature of its evaluations, producers of GE animals approach the agency as early in the development process as possible.

II. Product Definition

FDA and Aqua Bounty Technologies, Inc. (ABT) agreed to a Product Definition for AquAdvantage Salmon containing information regarding product identity, the demonstrated claim for the product, and limitations for the use of AquAdvantage Salmon. That Product Definition follows:

Product Definition for AquAdvantage Salmon

Product Identity

Triploid hemizygous, all-female Atlantic salmon (*Salmo salar*) bearing a single copy of the α-form of the *opAFP-GHc2* rDNA construct at the α-locus in the EO-1α lineage.

Claim

Significantly more of these Atlantic salmon grow to at least 100 g within 2700 deg C days than their comparators.

Limitations for Use

These Atlantic salmon are produced as eyed-eggs for grow-out only in the FDA-approved physically-contained fresh water culture facility.

III. Molecular Characterization of the Construct

A. Overview

Risk evaluation in the Molecular Characterization of the Construct step of the Hierarchical Review Process was essentially limited to characterizing the potential hazard(s) the rDNA construct might pose. In particular, we evaluated the intrinsic properties of the rDNA construct that might cause harm. The properties that were of most interest in this respect included potentially mobilizeable DNA sequences, or sequences encoding pathogens, toxins (including allergens), or substances likely to perturb the growth control of cells, tissues, or organs, except by explicit design. We also evaluated the purity of the construct in order to determine that unknown sequences were not introduced into the genetically engineered animals.

This evaluation was conducted in accordance with the approach later summarized in Guidance for Industry #187 *Regulation of Genetically Engineered Animals Containing Heritable Recombinant DNA Constructs*. As summarized in that document, typically, the information required for this section should include, but not be limited to, the following:

- a description of the source(s) of the various functional components of the construct;
- the sequence of the rDNA construct;
- the purpose of the modification;
- details of how the rDNA construct was assembled;

- the intended function(s) of the introduced DNA; and
- the purity of the preparation containing the rDNA construct prior to introduction into recipient animals or cells.

The evaluation of the Molecular Characterization of the Construct for AquAdvantage Salmon was organized in five general sections that address the characterization topics listed above.

1. Source and description of DNA for the inserted construct;
2. Construction, including method and intermediate organisms
3. Sequence of the final product.
4. Demonstration of promoter function in salmonid cells; and
5. Components of microinjection syringe

The available materials described construction and confirmation of a fish growth regulator (Chinook salmon growth hormone) under the control of transcriptional regulatory elements derived from ocean pout and Chinook salmon carried in a standard plasmid backbone. The constructs did not contain coding regions clearly derived from known toxins, pathogens, oncogenes, tumor suppressor genes, or sequences derived from transposable elements or retroviruses that would confer transgene mobilization. Thus, the evaluation of subsequent portions will focus on the Chinook salmon growth hormone gene and gene product, the ocean pout and Chinook salmon-derived regulatory sequences and the bacterial plasmid backbone.

We conclude that the data submitted are acceptable for the Molecular Characterization of the Construct portion of the hierarchical review of a new animal drug application for AquAdvantage Salmon.

B. Evaluation

1. Source and description of DNA

a. Plasmids

Although the plasmid portion of the rDNA construct is generally not intended to be inserted into the GE animal, an understanding of what plasmids were used to generate various intermediates in the assembly of the rDNA constructs was informative as to potential hazards associated with the plasmids as well as what rDNA to look for in subsequent steps.

Several closely related, and commonly used bacterial plasmids (pUC9, pUC13, pUC14 and pUC18) were used to generate various intermediates in the assembly of the rDNA construct used for generation of the AquAdvantage Salmon.

AquAdvantage Salmon were tested for pUC origin plasmid DNA sequences (see Molecular Characterization of the GE Animal Lineage, Section IV, below). No unanticipated sequences from these plasmids were found in the EO-1α lineage AquAdvantage Salmon.

b. Virus or Bacteriophage

No bacterial or eukaryotic viruses or sequences were used that would result in viruses or viral sequences being transferred to, or propagated in, the eventual GE animal.

c. Inserts

Recombinant DNA inserts from three sources were used for the final construct. These sources included regulatory sequences from ocean pout, the growth hormone coding region from Chinook salmon and small synthetic linkers to aid in assembly of the inserts and plasmid. This final construction is discussed in detail below in section 2b.

i. *Ocean Pout Anti-Freeze Protein (opAFP) Regulatory Sequences*
The upstream (5') and downstream (3') regulatory sequences used in the construct were obtained from a genomic isolate of a Type III anti-freeze protein (AFP) gene from the ocean pout (op). Isolation of the opAFP gene is available in J. Biol. Chem., vol. 263(24)12049-12055 (Hew et al., 1988). More information regarding the isolation of the opAFP regulatory regions can be found in Mol. Marine Biol. Biotech., vol. 1(4/5)290-300 (Du et al., 1992b).

ii. *Chinook Salmon Growth Hormone (GH) Coding Sequence*
The Chinook salmon GH gene was identified and isolated from a cDNA library prepared using pituitary gland of Chinook salmon. This cDNA is full-length and encodes a single, mature hormone.

iii. *Synthetic linkers*
Two synthetic DNAs corresponding to the 5' untranslated regions (UTRs) were prepared using established sequences of the Chinook salmon GH-1 and the ocean pout AFP. These dsDNA strands included 5' Bgl II and 3' Pst I sites, giving rise to 75 bp and 74 bp 5' UTRs, respectively. The GH-1 UTR was used for assembly of the opAFP-GHc construct, whereas the AFP UTR was used for the opAFP-GHc2 construct. This difference in 5' UTR constitutes the only reported difference between opAFP-GHc and opAFP-GHc2.

Conclusion: The submitted materials described a standard plasmid backbone, transcriptional regulatory elements derived from ocean pout, a fish protein growth regulator (Chinook salmon growth hormone), and synthetic primers. The material provided in the submissions did not suggest that there were any sequence elements in the constructs that contained coding regions clearly derived from known toxins, pathogens, oncogenes, tumor suppressor genes, or sequences derived from transposable elements or retroviruses that would confer transgene mobilization. Thus, the evaluation will focus on the Chinook salmon growth hormone gene and gene product, the ocean pout and Chinook salmon-derived regulatory sequences and the bacterial plasmid backbone.

2. Construction, including method and intermediate organisms

a. Assembly of the opAFP-GHc2 Construct

The assembly strategy used for the growth hormone construct used in AquAdvantage Salmon as well as constructs containing chloramphenicol acetyltransferase gene (CAT; used below to test promoter function in salmonids cells) is presented in the diagram provided below. Further, CAT is not a concern as it was not present in the final verified construct.

The multi-step assembly was typical for the time that the construct was assembled and utilized routine rDNA procedures. As described in section 3 below and in the Molecular Characterization of the GE Animal Lineage, the final rDNA construct (shown at the bottom of Figure 1) was verified.

b. *Bacterial Hosts of Construct*
The plasmids were propagated in, and isolated from *E. coli* K12 strain DH5α. This is a widely used laboratory bacterial strain and is not of concern in this context.

c. *Eukaryotic Cells as Potential Hosts of Construct*
The constructs were not propagated or expanded in eukaryotic cells prior to transduction of the fish eggs.

Conclusion: The level of description of the recombinant DNA techniques were appropriate and often exceeded the minimum description acceptable for this portion of the evaluation. The construction strategy generally employs techniques routine for laboratories in the field. Most of the techniques were routine for the time at which the constructs were generated, were often briefly described, and were based on protocols cited either in the primary literature or standard laboratory manuals such as Molecular Cloning (Sambrook et al 1989).

3. Sequence of the final construct

The DNA sequence was determined for the "insert" portion of the construct, not including the plasmid backbone. The sequencing coverage was at least two-fold for the entire insert. Sequencing coverage of the GH gene and immediately adjacent control regions was eight- to ten-fold, more than sufficient for sequencing the final construct. Chromatograms were provided that contain clear, well defined peaks that were typically readable for greater than 500 bases, consistent with good quality reagents and substrates. Thus, the observed sequence presented was reliable. Finally, the deduced amino acid sequence of the open reading frame of the GH gene was consistent with published materials.

Conclusion: We conclude that the sequence determination submission is acceptable and that the information is sufficient to support molecular characterization of the construct.

4. Demonstration of promoter function in salmonid cells

A series of experiments was provided to demonstrate that the ocean pout antifreeze type III regulatory regions were functional and that the promoter (or small parts of it) was functional in appropriate salmonid cell types (Du et al., 1992b; Gong and Hew, 1993). This supports the proposed use of the rDNA construct at this stage of the evaluation.

5. Components of microinjection syringe

Linearized DNA dissolved in 2 – 3 nL of 0.9% NaCl in sterile water was used in the microinjection to produce the founder animals. The purity of the preparation containing the rDNA construct prior to introduction into recipient animals or cells was acceptable.

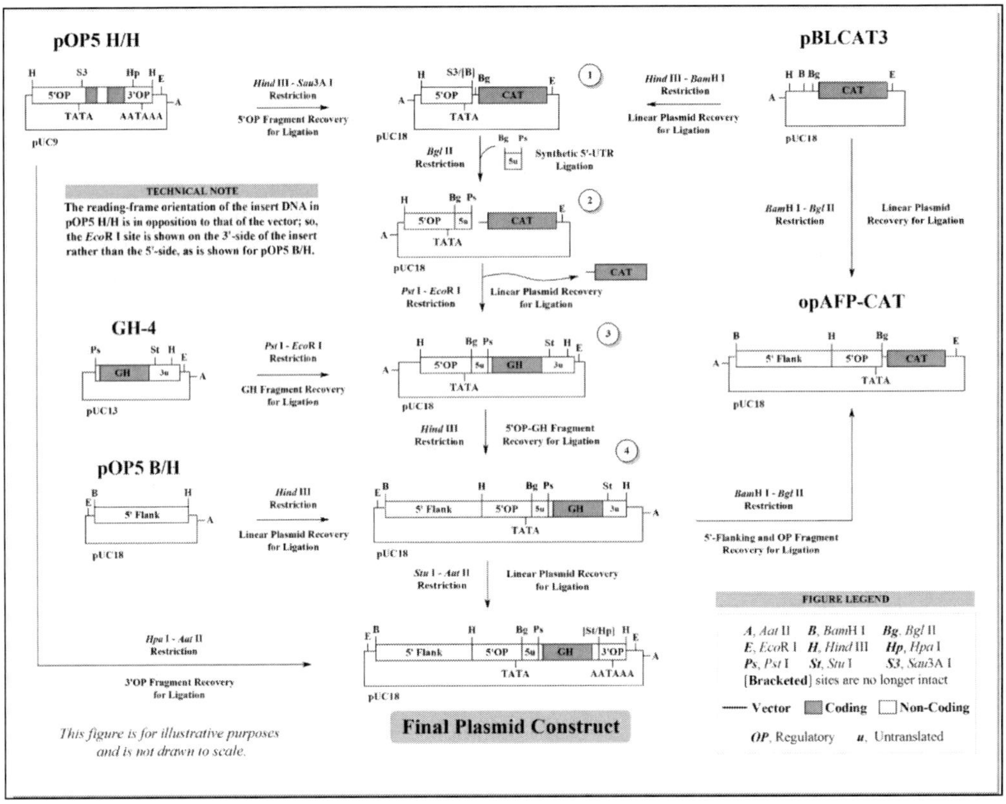

Figure 1. Schematic Summary of the Cloning Strategy Employed in Development of the opAFP-GHc and opAFP-GHc2 Constructs.

C. Conclusion

The general information provided by ABT as to the molecular construction of the vectors and transgenes injected is internally consistent. Supporting data for the sequence of the insert (but not plasmid) in the injected DNA is provided. Our evaluation of the submitted data do not identify any specific hazards intrinsic to the rDNA construct with the possible exception of the growth hormone gene that is present by explicit design.

We conclude that the data submitted are acceptable for the Molecular Characterization of the Construct portion of the hierarchical review of a new animal drug application for AquAdvantage Salmon.

IV. Molecular Characterization of the GE Animal Lineage

A. Overview

In this step of the hierarchical review process, we evaluated the molecular consequences of the insertion of the *opAFP-GHc2* construct into the EO-1α lineage of salmon. This evaluation was intended to identify any hazards that might result because of the integration event as well as the overall stability of the inserted construct in the lineage over multiple generations. Data in support of this step would generally include a full molecular characterization of the integrated construct at its site of insertion.

We evaluated the molecular characterization of the rDNA construct integrated in the genome of GE salmon. The rDNA construct is for expression of Chinook salmon growth hormone under the control of an ocean pout promoter. Although a number of lines of GE salmon were generated, ABT limited its production to specific lines within the EO-1α lineage. For this step of the hierarchical review process, we only considered data in support of the molecular characterization of the rDNA construct integrated in the GE animal.

There are two groups of hazard identification questions posed during this step.

With respect to the inserted sequences and their immediate flanking regions:

i. Does the GE animal contain sequences that are likely to pose potential hazards to the animal, humans, or animals consuming food from that animal, or the environment?
ii. Is the genotype changing over the life span of the animal or product?
iii. Is the inserted DNA what was expected from the data presented in support of the Molecular Characterization of the Construct?

In addition,

iv. Does the GE animal contain other contaminating or hazardous materials such as viruses, cells, or chemicals?

We conclude that the data submitted support the Molecular Characterization of the GE Animal Lineage portion of the hierarchical review of AquAdvantage Salmon. No hazards were identified, with the possible exception of the growth hormone gene itself, which is present by design, and will be evaluated at a subsequent step of the hierarchical review process.

B. Evaluation

1. Does the GE animal contain sequences that are likely to pose potential hazards to the animal, humans, or animals consuming food from that animal, or the environment?

To evaluate the consequences of the insertion of the rDNA sequence, we evaluated data and information characterizing (a) the number of insertion sites, and (b) the insertion site itself, including possible disruption of other genes and analysis of open reading frames (ORFs) within and around the insertion site

a. Number of Insertion Sites

Using information from the Molecular Characterization of the Construct, assays were designed and conducted to detect the rDNA construct and the pUC plasmid backbone sequences that could have been (but were not) inserted into the fish genome. This analysis addressed: (i) whether any plasmid DNA is present in the lineage; and (ii) the insertion sites present in initial, and production lines of fish. Three methods (Southern analysis, polymerase chain reaction (PCR) amplification, and DNA sequence analysis) were used to characterize the rDNA in the animal(s).

i. Plasmid DNA

The growth hormone expression construct was released from the bacterial pUC plasmid DNA prior to microinjection of the eggs but the pUC plasmid DNA was not removed from the mixture (see section 4 below), so it was necessary to determine whether it was present in the genome of the GE salmon. If the plasmid DNA was not present in the genome, then no assessment of hazard or risk associated with it was necessary. To determine if the pUC plasmid DNA was present in AquAdvantage Salmon, Southern analysis (with appropriate controls) was conducted. No pUC plasmid DNA was detected in any of the F_1 AquAdvantage Salmon descended from the EO-1α lineage. Notably, the fish in this study include the progenitors to all of the lines currently under development. If pUC plasmid was not present in these progenitors, then it could not be present in subsequent generations. Thus, we conclude that the pUC plasmid DNA is not present in the AquAdvantage Salmon lines being considered for approval.

Conclusion: pUC plasmid DNA was not inserted into the genomic DNA of the AquAdvantage Salmon lines being considered for approval and no further consideration of it is necessary.

ii. Number of insertion sites

Multiple experimental methods were used to characterize the number of construct insertion sites. These methods include Southern analysis, PCR amplification, and DNA sequence analysis. These analyses showed that although the initial GE animal contained two insertion events, the progeny lines developed for production contained one well-characterized construct at the α-locus.

Southern analysis demonstrated that early generations of AquAdvantage Salmon contained up to two distinct insertion sites, referred to as the α and β loci. The Southern analysis gave rise to multiple distinct bands that corresponded to these two copies of the *opAFP-GHc2* construct. ABT determined that the presence of the α- locus conferred the enhanced growth phenotype and fish containing the β-locus exhibited standard growth. Thus, ABT chose to select for the α-locus and bred the β-locus out of their production lines. Additional Southern analysis data was provided supporting the absence of the β-locus from the lines selected for production use (see section 2 below).

PCR amplification of the inserted construct in F_2 generation fish, followed by enzymatic digestion of the PCR products was a second method by which ABT determined the number of constructs integrated into the salmon genome. Primers were specifically selected to distinguish between: the α and β loci, the inserted growth hormone, native salmon growth hormone loci, and single insertion versus multiple insertions per site. These data were consistent with the Southern analysis discussed above. (Note: additional PCR analysis of the α-locus in F_2, F_4 and F_6 generation fish is discussed further below).

DNA sequence analysis was a third method employed by ABT to determine copy number and stability of the insert at the α-locus. Primers were designed to specifically anneal to the least conserved regions of the 5' and 3' genomic flanking regions around the α-locus. This allowed ABT to not only obtain better specificity in amplification of the construct, but also allowed ABT to monitor genomic stability over multiple generations. Eleven other primer pairs were designed to fully sequence the inserted construct at the α-locus. Sequence data provided were consistent with a single copy of the *opAFP-GHc2* construct at the α-locus. Sequencing of the α-locus was performed by a contract laboratory in accordance with Good Laboratory Practice standards (GLP; 21 CFR Part 58). Each sequencing reaction produced > 600 base pairs (bp) of good coverage. Chromatograms were provided and demonstrate the high quality of these data.

Conclusion: The AquAdvantage Salmon currently used for production contain a single well-characterized copy of the construct at the α-locus.

b. Evaluation of Insertion Site

Evaluation of the α-locus was aimed primarily at site specific effects including (i) disruption of genes at the insertion site and (ii) generation of a novel open reading frame by the recombination of the rDNA construct and genomic DNA.

i. Possible Disruption of Other Genes

The sequence data discussed above showed that the α insert was in a region of repeated DNA (a 35 base repeat). Thus, this insertion site was not a protein coding region. Additionally and importantly, if, as is expected with non-homologous recombination, part of the chromosomal DNA was deleted when the α insert was inserted, it is likely that only part of this 35 base repeat region was lost. Repeated regions like this are quite variable and nonessential, so loss of part of the repeat region is unlikely to adversely affect the fish.

Conclusion: The insertion of the construct at this site is not expected to impact the expression of native genes.

ii. Open Reading Frame (ORF) Analysis

As discussed above, the construct is located in a repeat region. Therefore, there are no open reading frames in the region flanking the insertion site and generation of novel open reading frames across the insert junction is not possible.

Furthermore, as discussed in the Molecular Characterization of the Construct (see Section III, above), the insert only contains sequences derived from Ocean Pout and Chinook salmon, both of which are commonly consumed as food. We have not identified any sequences that are of obvious concern; however, food consumption risks are evaluated during the Food Safety step of the hierarchical review (see Section VII, below).

Conclusion: Because the construct is inserted into a repeat region, there are no putative open reading frames other than those intended in the construct itself. Therefore, no additional risk from novel or altered open reading frames is present.

Overall Conclusion for Risk Question 1: With the possible exception of the growth hormone gene included by design, we have not identified any sequences that are likely to contain potential hazards to the target animal, humans, or animals consuming food from that animal, or the environment. ABT provided acceptable characterization of the α-locus.

2. Is the genotype changing over the life span of the animal or product?

The genotypic and phenotypic stability of the AquAdvantage Salmon have been assessed using a number of approaches. Several of these approaches utilize molecular biological analysis, including DNA sequence analysis, Southern analysis, and PCR analysis (described above). As discussed further below, we concluded from a "weight of evidence" analysis that the *opAFP-GHc2* construct is unchanged and stably maintained at the α locus over at least seven (7) generations (F_0 to F_6) and multiple lineages of AquAdvantage Salmon derived from the EO-1α founder.

The DNA sequence analysis of the α-locus was consistent (other than the rearrangement discussed below) from "the test tube" through one F_2 and one F_4 generation fish descended from EO-1α. The sequence of the coding region was unchanged in these samples. Additionally, the sequence of the genomic DNA flanking the α-locus (a repeated region) was also unchanged between the F_2 and F_4 generations in this line of fish. Thus, we conclude that the growth hormone expression insert is stably maintained at this chromosomal position (the α-locus).

Similar to the PCR analysis of F_2 generation fish described above, additional PCR analysis of the α-locus in F_2, F_4, and F_6 generation fish was conducted by ABT using GLP standards, and submitted for our evaluation. ABT developed detailed procedures and provided supporting data for a series of PCR amplifications that identified the *opAFP-GHc2* construct in the α locus. Specifically they described several primer sets and resulting amplification products corresponding to the 5' and 3' ends of the α-locus (with adjacent chromosomal flanking DNA) as well as the *opAFP-GHc2* growth hormone gene and two different endogenous growth hormone genes. The technique was appropriate, the method was well described, and the results were clear and unambiguous. ABT applied this method to samples from a total of 72 (74 if we include known positive and negative control fish) F_2, F_4 and F_6 generation fish well dispersed among the lineages currently being pursued for further development. This data was compelling evidence that the α-locus is stable over seven (7) generations.

To determine the stability of the α-locus over multiple generations as well as to confirm earlier PCR studies showing that the β-locus was successfully bred out of the later generation fish, additional Southern analysis was performed by a contract laboratory in accordance with GLP standards. Samples of 22 fish (including controls) dispersed among the lineages under development were subjected to Southern analysis. The analysis and some of the samples used replicate the Southern analysis described above. The results of the earlier study and those reported here were consistent. Here only the previously identified F_2 generation fish contained the α and β loci where all other F2, F4 and F6 generation fish contained only the α-locus. No negative controls contained an insert.

Conclusion: The α-locus is stable between the founder and the seventh generation (F_0 through F_6). The β-locus was selectively bred out of the lineage and is not present in the lines of fish currently in production.

3. Is the inserted DNA consistent with the data presented in support of the Molecular Characterization of the Construct?

The general structure of the growth hormone expression construct, stably maintained in ABT salmon[2], has been characterized and is consistent with data presented for the Molecular Characterization of the Construct evaluation. Characterization and analysis of the construct in the animal is provided (see also Transgenic Res 2006;15:465-480 (Yaskowiak et al., 2006)). This analysis identified a rearrangement compared with the original construct (Panel A Figure 2). The rearrangement displaced a portion of the far 5' non-coding regions of the insert to the 3' end of the insert (Panel B Figure 2). However, this rearrangement is not a concern as discussed further below.

ABT suggested that the rearrangement identified in the EO-1α lineage occurred during the initial transformation event that resulted in generation of the founder of this lineage of fish (EO-1α). They proposed that circularization of the *in vitro* linearized DNA prior to integration of the DNA into the fish chromosome resulted in a rearrangement of the elements in the original construct at the α-locus. This simple and plausible explanation provided a reasonable model for how the final molecular structure of the α-locus arose. The rearrangement moves the far upstream promoter regions (typically enhancer domains) to a downstream location relative to the growth hormone coding region. As detailed in the Molecular Characterization of the Construct evaluation (Section III, Part B.4), ABT previously provided data demonstrating that the far upstream regions of the promoter were not required for expression from this promoter. Furthermore, enhancer elements act at a distance and are generally not orientation dependent. This rearrangement is well characterized, is not of concern, and requires no further consideration in future evaluations.

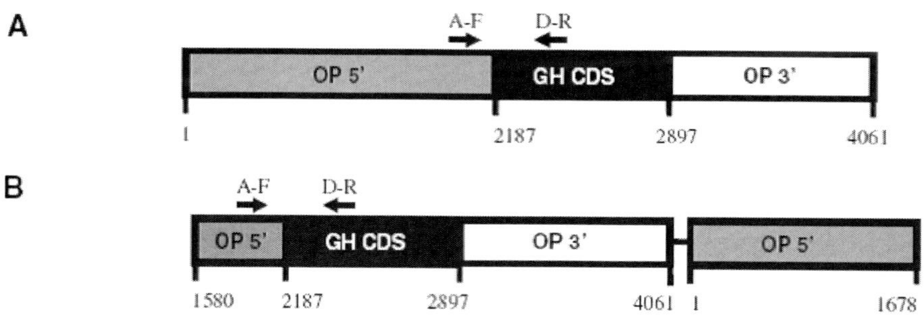

Figure 2. Characterization of the rDNA construct at the EO-1α locus. (A) Structure of the opAFP-GHc2 construct that was injected into Atlantic salmon eggs to produce the EO-1α strain of transgenic Atlantic salmon. (B) Schematic of the genomically integrated transgene EO-1α. (From Figure 3 in Hobbs and Fletcher, 2008.)

Conclusion: With the exception of a well-characterized rearrangement, the sequence of the integrated construct was consistent with the sequence of the rDNA construct. No additional evaluation, above that which would be normally conducted, was required in other steps. We note that the rearrangement may be useful in the future for identification as it is likely to be unique to this specific lineage.

4. Does the GE animal contain other contaminating or hazardous materials such as viruses, cells, or chemicals?

As discussed in detail below, no hazards have been introduced by the methods used to either prepare and purify the construct, or microinject the construct into the eggs.

Preparation and purification of the construct DNA for injection into salmon eggs was provided. The insert was excised from the bacterial plasmid sequences by overnight digestion with restriction endonuclease *Eco*RI, followed by phenol/chloroform extraction and ethanol precipitation. The insert was linearized and released from the plasmid, but not purified to remove the plasmid fragment from the solution. Thus, the GH containing inserts from the plasmids as well as the pUC plasmid sequences were microinjected into the eggs. (As described above in Part A.1.a.i above, pUC plasmid DNA was not incorporated into the genome of the EO-1α lineage salmon being developed for production and so is not a concern.) The descriptions of the molecular biological methods and procedures included in the submission are typical of standard procedures routinely used at the time the work was conducted.

Microinjection of salmon eggs is described in several submissions. The description is acceptable. Several publications (Du et al., 1992a; Du et al., 1992b; Fletcher et al., 2001; Shears et al., 1991) support the conclusion that the methods used in the production of AquAdvantage Salmon are consistent with the methodology generally in use at the time of injection.

Conclusion: There is no risk from any contaminants or other hazardous materials (excluding the possibility of growth hormone) in the EO-1α lineage.

C. Conclusion

The information provided by ABT as to the Molecular Characterization of the GE Animal Lineage is consistent and in agreement with the Molecular Characterization of the Construct and is acceptable. Supporting data for the sequence of the injected rDNA construct and the molecular stability of the construct over seven generations have been provided.

We conclude that the data submitted support the Molecular Characterization of the GE Animal Lineage portion of the hierarchical review of AquAdvantage Salmon. No hazards were identified, with the possible exception of the growth hormone gene itself, which is present by design, and will be evaluated at subsequent steps of the hierarchical review process.

V. PHENOTYPIC CHARACTERIZATION

A. Overview

This section evaluates the phenotypic characterization of GE salmon containing a construct for expression of Chinook salmon growth hormone under the control of an ocean pout promoter ("AquAdvantage Salmon"). Although a number of lines of GE salmon were generated, ABT has limited its production to specific lines derived from the EO-1α founder. The information evaluated often contained data in support of other levels of the hierarchical review process (e.g., durability, claim validation); however, in the context of this evaluation, the primary focus was on the adequacy of data to support the phenotypic characterization of the GE animal, and to draw conclusions regarding animal health.

To this end, we conducted a weight-of-evidence evaluation of the AquAdvantage Salmon phenotype. After evaluating all of the data in the submissions, and referring to peer-reviewed publications as appropriate, we identified no significant hazards or risks with respect to the phenotype of AquAdvantage Salmon as a result of the *opAFP-GHc2* construct (the AquAdvantage construct). There are no significant adverse outcomes associated with the introduction of the AquAdvantage construct and the production of triploid monosex GE Atlantic salmon. Most of the adverse outcomes that have been observed (e.g., morphological changes) were present in comparators or have been described in the peer-reviewed literature with attribution either to the induction of triploidy or to non-transgenic rapid growth phenotypes. Most of these adverse outcomes occur in the early life stages; their consequences are likely to be small and within the range of abnormalities affecting rapid growth phenotypes of Atlantic salmon. None of the adverse outcomes noted, which were minimal, are expected to have any implications for food consumption risks; some may affect the fitness of GE animals such that any escapees from containment would be less capable of surviving.

As with all data sets, there are some uncertainties. The primary area of uncertainty is determining the actual rate of adverse outcomes in grow-out facilities, as the relatively heavy culling rate that occurred in the space-limited broodstock facility described in these data sets may have influenced the apparent rate of abnormalities. In order to gather more information in actual grow-out conditions, CVM has recommended a surveillance program as part of the durability plan (see Genotypic and Phenotypic Durability Plan, Section VI below), and will closely monitor post-market surveillance reports of adverse events.

We applied a risk-based approach to evaluate the GE animals to address four risk/hazard questions developed for the phenotypic characterization of GE animals:

1. Is there direct or indirect toxicity to the animal?
2. Are there phenotypic characteristics that identify hazards for other steps in the evaluation?
3. What are the risks to the user (user safety)?
4. What are the risks to the animal from any components of any biological containment strategy?

B. Evaluation

In our weight-of-evidence evaluation of the AquAdvantage Salmon phenotype, we draw on data from a number of sources. These include the following, listed in rank order (from highest to lowest) of importance in the overall weight-of-evidence evaluation: (1) controlled studies conducted on the specific animals being considered for approval; (2) other non-controlled studies on these same animals; (3) historical hatchery records and data for these animals; and (4) studies reported in the scientific literature investigating these same animals or their relatives. Each source, in turn, is given appropriate deference with respect to its relevance to the risk or hazard identification question under consideration.

To evaluate the phenotypic consequences of the insertion of the AquAdvantage construct, we evaluated (a) the general husbandry conditions for the fish; (b) the specific conditions at the current facilities; (c) any general health observations; (d) feed consumption and weight gain rates, (e) overall mortality and morbidity; (f) physical exams; (g) clinical pathology assessments (clinical blood counts and chemistry panels); (h) necropsies (post-mortem examinations including histopathology); (i) disease resistance; (j) smoltification and seawater survival; and (k) several other phenotypic characteristics for which data were available. Nutritional (compositional) and hormonal data for adult AquAdvantage Salmon were considered as part of the FDA evaluation of food and feed safety (see Food Safety, Section VII below).

a. General Husbandry Conditions

Because fish husbandry conditions, particularly those that affect water quality, can affect fish health and phenotype (e.g., morbidity, mortality and stress-related parameters), we included in our assessment of phenotype a consideration of husbandry conditions. AquAdvantage and comparator[3] salmon have been cultured at ABT's Prince Edward Island (PEI) facility under standard conditions for the freshwater (hatchery and smolt production) phase of salmon aquaculture[4]. Water specifications are maintained at the following conditions: pH = 7.3; oxygen > 8 mg/L (range of 11.7 – 17.7 mg/L); carbon dioxide < 20 ppm; ammonia < 0.03 ppm; nitrate < 40 ppm; nitrite < 0.15 ppm; and stocking densities of 10-35 kg/m^3. Temperature ranges were from 12.1 – 14.3°C. Corrective action is taken to bring water quality within these parameters, if needed.

Fish are generally fed to satiety using a commercial salmon diet of appropriate composition and pellet size. Records are maintained for water quality parameters and tank feed amounts.

Conclusion: The general husbandry conditions for AquAdvantage Salmon are consistent with commercial freshwater aquaculture conditions; they present no identifiable hazards or safety concerns.

b. Specific Facility Conditions

The PEI facility is an aquaculture facility almost entirely dedicated to hatchery operations. The facility is licensed by the Canadian Department of Fisheries and Oceans (DFO) and is certified as disease-free under Schedule II of the Canadian Fish Health Protection Regulations (FHPR), C.R.C., c. 812. Schedule II pathogens include, among others, those that cause: viral hemorrhagic septicemia (Egtved virus, VHSV), infectious hematopoietic necrosis (IHNV), infectious pancreatic necrosis (IPNV), whirling disease (*Myxobolus cerebralis*), ceratomyxosis (*Ceratomyxa shasta*), furunculosis (*Aeromonas salmonicida*), and enteric redmouth disease (*Yersinia ruckeri*).

The large tanks at the PEI facility are 11.2 m^3 cylindrical tanks using recirculating ground water. The water is adjusted, as necessary, to meet the specific water conditions described previously.

The fish stocking density during the study was generally within the range of 10 – 35 kg/m^3, a range representative of commercial salmon aquaculture conditions.

The ABT broodstock facility and available records were inspected by representatives of CVM and FDA's Office of Regulatory Affairs. The inspection found the facility acceptable, no FDA Form 483 was issued[5], and the inspection was classified as "No Action Indicated" or NAI.

A similar site visit of the ABT Panama facility was conducted by staff from CVM and the National Oceanic and Atmospheric Administration. This site visit was conducted primarily to verify that there was acceptable physical containment at the facility. In addition, the water quality and rearing conditions at the facility and the general health of the fish in residence were also examined. Nothing was observed that would indicate an issue of concern with respect to the facility or the fish therein.

It was noted that the culture conditions (e.g., water temperature, pH, alkalinity, etc.) were likely to be significantly different from the facility at PEI as a result of differences in, among others, water source, facility design, and environmental factors due to geographic location. Although within the range of culture conditions for Atlantic salmon, the effect of the differences between the PEI and Panama facilities, especially temperature, on the resulting AquAdvantage phenotype is unknown.

Conclusion: The husbandry and rearing conditions at the PEI and Panama facilities do not present specific concerns with respect to animal health.

c. General Observations

ABT conducted a controlled, blinded animal safety study at their PEI facility, which examined the effect of the integrated AquAdvantage construct on the health of the GE animals. The study was a careful and well-controlled assessment of the health of the fish and the following sections focus primarily on the information provided in this study.

The general health and behavior of all of the fish was assessed by facility personnel and independent veterinary professionals specializing in aquatic species. The study fish were observed at the PEI facility at four separate time points following pre-enrollment qualification of the fish from each study group. Assessments were made for feeding activity, behavior, posture, and position in the water column, coloration, observation of any external lesions, morbidity, mortality, and any other abnormal clinical signs. No health abnormalities were observed and the fish were regarded as in good health and of normal behavior.

Conclusion: AquAdvantage Salmon show no general health or behavioral abnormalities relative to comparator fish.

d. Size, Weight, and Related Parameters

Understanding the design and interpreting the results of the study requires accommodating both the effects of the intended rapid early life stage growth of the GE fish, and general salmon spawning behavior. AquAdvantage Salmon are intended to grow more quickly in early life than their comparators. This different growth rate results in harvesting fish at either the same age but different sizes, or at about the same size (e.g., "market size"), but at different ages. Because salmon spawning is seasonal, in the latter case, this implies harvesting at different times of year.

The design used for the study incorporated harvesting the animals in the main comparison groups when the individual fish reached 1,000 – 1,500 g; these animals will have experienced different growth conditions up to, and at the time of, harvest. For example, fish harvested in February will likely be experiencing shorter days and exposed to colder water with potentially lower microbe burdens than fish harvested in late summer. To address the differences in growth conditions, the study design included a smaller group of "satellite control" fish (non-GE fish of respective ploidy), referred to as SAT or satellite controls, which were harvested at the same time as the AquAdvantage Salmon, and are therefore age-matched comparators.

As previously discussed, all fish in the animal safety study and, in fact, all fish at ABT's facilities, are fed to satiety using appropriate size-adjusted feeding rates of standard salmon feed, as is standard practice for the aquaculture industry.

At least six fish per gender in each of four groups were enrolled in the study: diploid AquAdvantage; triploid AquAdvantage; diploid comparators; and triploid comparators. A total of 48 fish were enrolled in the final study from between 400 and 800 candidate fish (100-200 for each group). Once the fish reached between 15 - 30 g in weight, each candidate fish was given a passive integrated transponder (PIT tag) and a unique fish identification number (UFID) for identification. During the prequalification phase, fish showing clear signs of morbidity were excluded.

Fish enrolled in the animal safety study were evaluated for size by measuring overall body weight, fork length, condition factor, and gonadosomatic index. Statistical analysis of the data showed no differences among study groups with one exception. The body weight of diploid and triploid AquAdvantage Salmon was much greater than that of corresponding satellite controls. This was expected given that the construct was intended to result in accelerated early life-stage growth.

Conclusion: There are no adverse effects on size, body weight, or related parameters in AquAdvantage Salmon relative to comparator fish other than the effects expected from the introduction of the AquAdvantage construct.

e. Physical Examinations: Behavioral and Physical Abnormalities

Specific physical evaluation of adult (1,000-1,500 g) fish enrolled in the animal safety study occurred on three distinct levels. First, fish were observed during the study for avoidance and feeding behavior, posture-position in the water column, and any other observed behavioral or physical abnormalities. Throughout the pre-enrollment phase, no abnormal behavior was observed for either the AquAdvantage or the comparator groups.

Second, a gross external examination was conducted on each fish enrolled in the study. This exam included nine specific observations, and included photographs of each fish. Individual fish were then given a rank score where Rank 1 was no change versus a perfectly-formed Atlantic salmon; Rank 2 was a slight change; Rank 3, a moderate change typical of farmed salmon; and Rank 4 was a severe change that could affect commercial viability or fitness. The results are presented in Table 2.

These results appear to indicate that the occurrence of external abnormalities was similar, if not lower, in AquAdvantage Salmon versus comparator salmon and suggest that the induction of triploidy, not the introduction of the AquAdvantage construct *per se*, accounts for the differences in abnormal findings.

The study had a design limitation that could potentially affect the results and thus our interpretation. The design limitation results from the high rate of removal of early-life stage fish (e.g., fry or smolts). ABT indicated that this practice normally occurred at the ABT PEI hatchery at regular intervals because of space limitations. The net result is that the adult fish in the study may not reflect the nature or incidence of abnormalities of the initial population.

According to the information ABT provided to us, *ad hoc* culling was historically effected in association with inventory management activities (i.e., removing excess inventory, biomass reduction, separation of fast-growing individuals from slow-growing individuals, and broodstock selection) and often no data were collected on fish culled as excess inventory, particularly early life stages (i.e., eggs, yolk-sac fry and first-feeding fry). Although not specifically described in the report of this study, one would reasonably expect that the culling was done in a manner that selected for improving the broodstock, thus retaining the healthiest and fastest growing individuals in the facility. We have no reason to believe that ABT's culling practices were inconsistent with the approaches used in broodstock operations in the commercial salmon industry; these may differ from commercial grow-out facilities.

Abnormalities are known to be induced by triploidy in Atlantic salmon (see further discussion below); these abnormalities are often observed in very young fish soon after hatching. Thus, although the removal of abnormal fish through selective culling, particularly those with moderate to severe abnormalities, may have skewed the population of adult fish available for inclusion in this study, the study results nonetheless are likely representative of those that occur in a commercial situation with continual culling of abnormal fish.

Additional data addressing the appearance of abnormalities in adult salmon are available from the AquAdvantage animal safety study. Results for morphological rankings of external appearance conducted during the pre-qualification phase of this study are summarized in Table 3 below. In this study, the triploid AquAdvantage Salmon group had the lowest total frequency of morphological changes (10.2%), while triploid non- transgenic salmon had the highest total percentage of malformations (23.0%). These results must also be interpreted carefully given our previously-mentioned concerns with continuous culling at the hatchery. Although these culling practices were not documented in the study report, we assume that such culling occurred because a related the report indicates that the number of fish within a rearing tank is typically reduced by planned culling before the fish reach tagging size.

Table 2. Results of Gross External Examinations

	Number of Abnormal Findings					
Feature	***ABT Salmon***		***Comparators***		***Satellite Controls***	
	Diploid	***Triploid***	***Diploid***	***Triploid***	***Diploid***	***Triploid***
Jaw	1	1	0	1	0	0
Operculum	2	0	0	0	0	0
Gills	0	10	4	12	0	4
Fin Structure	0	3	1	3	0	1
Vertebral Column	1	0	1	1	0	1
Eyes-Cornea	0	0	1	0	1	0
Skin	0	0	0	0	0	0
Color-Markings	0	0	0	0	0	0
Other: Cranium	0	0	1	0	0	0
Total Findings	4	14	8	17	1	6

Table 2. (Continued)

Feature	Number of Abnormal Findings					
	ABT Salmon		*Comparators*		*Satellite Controls*	
	Diploid	*Triploid*	*Diploid*	*Triploid*	*Diploid*	*Triploid*
Fish Without Findings	9	2	6	0	5	2
Number of Fish	12	12	12	12	6	6
Rank 1	9	2	6	0	5	2
Rank 2	3	10	6	11	1	4
Rank 3	0	0	0	1	0	0
Rank 4	0	0	0	0	0	0

Table 3. Percentage of Scores by Rank for External Appearance in the Phase I Pre-Qualification Selection for Salmon in the Animal Safety Study (2007 Year-Class)

Ploidy & Group	**ABT Salmon**					**Non-GE Salmon**				
	N	**Rank**				**N**	**Rank**			
2n		**1**	**2**	**3**	**4**		**1**	**2**	**3**	**4**
Included	97	82.5	16.5	0	1.0	94	88.3	11.7	0	0
Excluded	97	80.4	15.5	4.1	0	93	78.5	21.5	0	0
Total	194	81.4	16.0	2.1	0.5	187	83.4	16.6	0	0
3n	**N**	**1**	**2**	**3**	**4**	**N**	**1**	**2**	**3**	**4**
Included	85	90.6	8.2	0	1.2	49	59.2	28.6	8.2	4.1
Excluded	82	89.0	6.1	4.9	0	48	75.0	8.3	10.4	6.3
Total	167	89.8	7.2	2.4	0.6	97	67.0	18.6	9.3	5.2

Ranking scale for morphologic changes: 1 = none; 2 = slight; 3 = moderate; 4 = severe
Note: determination of inclusion or exclusion was made by random selection
2n = diploid; 3n = triploid

According to ABT, in some cases, the non-GE lower mode siblings of a cross are culled; alternatively, a predetermined number of fish are netted out and culled, or smaller or fish with irregularities are culled. For the safety study, it is not known whether culling was comparable for all four study groups, but, in general, this practice would be expected to remove those fish with moderate to severe malformations from all sample populations well before actual enrollment in the study began. This may explain why most morphological changes in the study, independent of the study group examined, were classified as "slight" in nature. Slight irregularities are less recognizable than severe ones, thus culling of these fish would have been less likely to occur. However, even if the irregularities are recognizable, culling of the fish with slight abnormalities would not be expected to be a common practice in a commercial setting because of the negative economic consequences of doing so. It is not unlikely, however, that such culling could occur in broodstock facilities.

In addition to the well controlled study described, ABT provided a summary of historical data addressing the health of AquAdvantage Salmon in several consecutive year-classes. Morphologic ranking data were reported for diploid and triploid fish of the 2003-2007 year-classes and are summarized in Table 4 below. These reports did not describe how these historical data were compiled or whether they included assessments of culled fish. Therefore, the inferential value of these data is limited and may be subject to the same concerns described previously for the safety study.

Nine internal organs were examined post-mortem in the animal safety study. These organs and structures were determined to be either normal or abnormal, and samples were taken for histopathology and other microscopic testing of abnormal findings. The weight and size index (the ratio of organ weight to body weight) was determined for the gastrointestinal tract, heart, liver, and gall bladder. No differences were found between AquAdvantage Salmon and their appropriate age- or size-matched comparators.

Gross morphologic and animal health data from earlier generations of diploid and triploid AquAdvantage Salmon and non-GE comparators were reviewed.

Table 4 summarizes the adverse outcomes noted for diploid and triploid AquAdvantage and comparator Atlantic salmon. The data indicate that, in general, irregularities decrease over time, in both rate and severity for the AquAdvantage diploid and triploid fish. With the exception of the 2006 and 2007 year classes, this trend is also observed in the comparator salmon. Most of the irregularities observed were classified as “2”, slight to moderate. The 2005 year class is an exception to these observations, and is discussed below.

Triploidy has been associated with an increased level of abnormalities in Atlantic salmon, typically characteristic lower jaw malformations (Benfey, 2001; O'Flynn et al., 1997). Although there are more (>30%) slight to moderate irregularities in triploid AquAdavantage Salmon than in their non-GE comparators in three of the five year- classes, similar results were observed in the diploid ABT salmon and comparator salmon. Therefore, induction of triploidy may not be the sole causative agent for the increased irregularities.

Examination of the rate of irregularities over time indicates that there was a large reduction in the percentage of slight-moderate abnormalities in the triploid AquAdvantage Salmon. In 2004, modifications were made to the PEI facility intended to improve conditions and animal health, which likely contributed to the improved morphology of all of the animals. For example, slight-to-moderate irregularities in the 2007 year-class, the most recent for which data is available, were only 7.1%, substantially lower than the range of 42.1 to 61.0% found in the 2003 to 2005 year- classes.

Interestingly, the decrease in irregularities was not as notable in non-GE comparators. For example, the incidence of irregularities in triploid non-GE salmon was highest in the 2006 and 2007 year-classes (33.1% and 71.5%, respectively). This high rate is likely not due to the induction of triploidy alone (in which case the rate of irregularities in triploid AquAdvantage Salmon should also have been elevated). Rather, it may be a function of the underlying genetics of the broodstock families used in the breeding crosses, or possibly, other factors. A family (genotype) effect has previously been observed on survival and other performance measures of diploid and triploid Chinook salmon (Johnson et al., 2004).

The 2005 year class presents an abnormally high rate of irregularities in AquAdvantage Salmon relative to their non-GE comparators. The reasons for this are not entirely clear. The rates are based on a very small population size; for example, there were only 38 fish in the triploid population. In contrast, the sample size ranged from 183 to 1165 fish per year class for triploid AquAdvantage salmon. The number of diploid AquAdvantage Salmon in the 2005 year class was more consistent with other year classes and exhibited a lower rate of irregularities. It is possible, therefore, that the high rate of irregularities in triploids in the 2005 year class was simply due to the small number of animals. The concomitant increased rate of irregularities in the diploid cohort, albeit at a lower frequency, does not provide a full explanation for the increase.

Further examination of the entire data set indicated that with the exception of the 2005 year-class, non-marketable severe irregularities have not been demonstrably higher in AquAdvantage Salmon compared with non-GE comparators, making the 2005 year class an outlier. ABT indicated that changes in incubation procedures may have been responsible for these effects, but did not provide a more detailed explanation. More recent summary data for the 2006 and 2007 year-classes (see also Table 4) are supportive of this contention and do not indicate elevated levels of severe abnormalities in diploid or triploid AquAdvantage Salmon compared to either diploid or triploid non-GE salmon. Therefore, the 2005 year class appears to be an outlier with respect to severe abnormalities. Although not common, as discussed further below, extreme rates of severe abnormalities in a given year class, cross, or geographic location have been reported in the literature (Sadler et al., 2001).

Table 4. Percentage of Irregularities by Rank in Diploid (2n) and Triploid (3n) Fish for the 2003-2007 Year-Classes of ABT Salmon and Non-GE Salmon

Ploidy & Year Class	ABT Salmon				Non-GE Salmon			
	N	Rank			N	Rank		
2n		**1**	**2**	**3**		**1**	**2**	**3**
2003	1327	42.1	57.9	0.1	215	94.4	5.6	0
2004	2368	91.7	8.1	0.2	627	100	0	0
2005	1586	17.2	70.4	12.5	816	98.7	1.0	0.4
2006	1276	61.4	36.1	2.4	544	97.8	2.2	0
2007	1792	95.2	4.3	0.5	664	85.1	14.4	0.6
3n	**N**	**1**	**2**	**3**	**N**	**1**	**2**	**3**
2003	1165	39.1	59.3	1.6	233	80.7	19.3	0
2004	328	36.0	61.0	3.1	92	96.7	1.1	2.2
2005	38	7.9	42.1	50	82	89.0	9.8	1.2
2006	289	72.3	27.0	0.7	148	66.2	33.1	0.7
2007	183	92.4	7.1	0.6	193	28.5	71.5	0

Ranking scale: 1 = no irregularity; 2 = slight-moderate irregularity; 3 = severe irregularity

Note: Scoring for the 2007 year-class used a 4 point scale rather than the 3 point scale used earlier. In order to make the results comparable all year-classes, totals for fish with slight and moderate irregularities (Ranks 2 and 3) were combined for the 2007 year-class.

ABT also provided a short white paper, prepared in response to a FDA request, addressing the occurrence and origin of morphological irregularities in salmonids and summarizing data presented in several other submissions on abnormalities found in AquAdvantage Salmon. The section of the white paper discussing the frequency and etiology of malformations in salmonids is quite short (~1 page) and provides little more than an overview of the issues and a citation of some of the relevant scientific literature. We performed additional literature searches on specific topics, for example, the frequency of malformations in commercial salmon. The discussion below reflects information from the white paper and from this additional literature search.

According to ABT's white paper, many factors and/or conditions have been associated with developmental abnormalities in salmon, including deficiencies in phosphorus and vitamin C, excess vitamin A, high or variable temperatures during early growth phases, exposures to certain drugs (e.g., oxytetracycline), contaminants in feeds (e.g., heavy metals, insecticides, PCBs), and some parasites (Vågsholm, 1998). Skeletal and jaw malformations are reportedly quite common (up to 80%, as discussed below) in salmon and trout reared on commercial farms, and may result in decreased productivity due to decreased survival, growth or consumer rejection. Supporting the ubiquity of observation of skeletal abnormalities in farmed Atlantic salmon, a recent study proposes a classification system that describes 20 different types of vertebral column malformations in Atlantic salmon that are repetitively observed under farming conditions (Witten et al., 2009).

The frequency of deformities in farmed fish seems to vary widely depending on fish genetics, local husbandry conditions, level of examination given and other factors. According ABT's white paper, veterinary field studies have identified the periodic occurrence of spinal compression (humpback) in 70% of salmon in Norwegian farming operations (Kvellestad et al., 2000) and jaw malformations in 80% of salmon at commercial sites in Chile (Roberts et al., 2001). Published data on commercial farming operations are not widely available and fish farmers are not generally open to sharing this type of information. Citing the study of Ørnsrud and Waagbø (2004), the white paper states that for general reference, the background occurrence of malformations in fish used as controls in various studies is generally less than 5%. This appears to be a reasonable gross estimate of what might be expected in wild populations and cultured populations not subject to disease or environmental stressors (e.g., poor water quality, contaminants or nutritional deficiencies). Therefore, a background rate of malformations of approximately 5% would not be unexpected.

One recent study in Norway found a frequency rate of deformed Atlantic salmon (percent of individuals with one or more deformed vertebrae) that ranged from 6.6% to 17.1% (Fjelldal et al., 2009). In the Fjelldal *et al.* study, neither genetic background, smolt quality, or off season smoltification was found to be an important factor in the etiology of vertebral malformations in farmed Atlantic salmon. In contrast, a recent study of Chinook salmon (Evans and Neff, 2009) found a very high variability in the overall frequency of spinal deformities between different families within the same fish population, with spinal deformities affecting up to 21% of the offspring within susceptible families; however, the overall frequency of malformations when looking across families in the two fish populations that were examined was less than 1%.

Certain abnormalities seem to be associated with the induction of triploidy. In his review article on the physiology and behavior of triploid fishes, Benfey (1999) states that although in general, triploids have similar, if not identical morphological and meristic characteristics as diploids, several specific morphological differences and abnormalities have been associated with triploidy in fish. The best and most frequently described gross anatomical abnormality in triploid fish is the lower jaw deformity in triploid Atlantic salmon, which may be linked to rapid growth rates in seawater. The two other abnormalities described in triploid Atlantic salmon, cataracts and changes in erythrocyte size, may be due to nutritional deficiencies. Sadler *et al.* (2001) has also described a gill filament deformity syndrome which is found at a much higher frequency in triploids (in up to 60% of triploid smolts) than in diploids.

There is some controversy in the literature as to whether it is the triploid condition itself, or the process by which it is induced (e.g., pressure or heat shock) which causes abnormalities in fish (Piferrer et al., 2009). Evidence for both causes has been presented and, in many cases, it has been impossible to separate the effects of the two. Even the specific process by which triploidy is produced may have an effect. For example, in one recent study with rainbow trout, Haffray *et al.,* (2007) found that triploidy induced by temperature shock produced morphological anomalies in fry at a higher rate than triploidy produced by pressure shock (11.7% vs. 2.8%), which in turn produced abnormalities at a rate not much different from that in diploids (1.9%). (Pressure shock is the method used to produce triploid AquAdvantage Salmon.)

Conclusion with respect to physical and behavioral abnormalities: Analyses of the behavior and gross external abnormalities of market size (1,000 – 1,500 g) AquAdvantage Salmon show no demonstrable differences from the comparator fish population when reared under growth conditions in ABT's PEI facility.

Although we have no reason to believe that ABT's culling practices are inconsistent with the approach used for broodstock development in the commercial salmon industry, culling procedures at the PEI facility are not likely representative of those used in commercial production and grow-out settings. Consequently, there is some uncertainty regarding the likelihood or incidence of abnormalities of AquAdvantage Salmon under commercial rearing conditions. To this end, the Durability Plan includes monitoring, data collection, and reporting of abnormalities observed under commercial production and grow-out conditions at the Panama facility where AquAdvantage Salmon will be reared.

f. Overall Mortality and Morbidity

During the ten-month animal safety study, 25 of 400 to 800 candidate fish were culled due to non-viability, morbidity, or mortality. Ten AquAdvantage Salmon (3 diploid and 7 triploid) and fifteen size-matched comparators (10 diploid and 5 triploid) comprised these "for-cause" removals from the study, for an overall removal rate of 6.25%. Numbers were similar between AquAdvantage and comparator fish as well as between diploids and triploids.

Of the 25 culled fish, 22 were subjected to histopathological analysis[6]. The analysis of these 22 fish showed small inflammatory changes in both AquAdvantage and comparator salmon. These changes were regarded as normal and typical findings in Atlantic salmon in aquaculture. No other abnormalities were identified.

Mortality observations at all life stages have been recorded at the PEI facility since 1996. Survival to first feeding varied significantly from year to year, and sometimes between different spawning crosses in the same year, but in general, survival at this stage was similar on average between AquAdvantage and non-GE salmon (Table 5). Low survival, when it occurred, was attributed primarily to fungi and opportunistic bacteria, and as a result, offspring of both GE and non-GE crosses periodically required treatment with drugs such as formalin, chloramine-T and salt. ABT reported that survival in the early rearing area improved since the facility upgraded to combi-tanks[7] in 2004; however, the survival data for the 2005 spawning year were not wholly consistent with this assertion.

Other researchers have found elevated mortality in triploid salmon prior to the start of first feeding (Benfey, 2001; O'Flynn et al., 1997)(Atlantic salmon) and early in development (Johnson et al., 2004)(Chinook salmon) relative to diploid comparators. This type of elevated mortality was not found in triploid AquAdvantage Salmon, possibly due to genetic selection in the diploid broodstock. This would be consistent with the findings of Johnson *et al.* (2004), whose results suggest that improvement of triploid performance (e.g., survival, growth) may be possible by selecting specific combinations of high-performance diploid broodstock for the production of triploids over several generations.

There are limitations in the interpretation of historical data. Because of space limitations, there was extensive culling at the ABT PEI facility which was often done on what is described as an *ad hoc* basis. For example, excess inventory of eggs and of early-life stages were removed at several different time periods: (1) between egg fertilization and hatching; (2) after hatching and before separation; (3) at the time fry were separated (≥5 g; e.g., when slow-growing fry were separated from fast growing fry); and (4) after PIT- tagging during the grow-out phase. Typically, no data were collected on fish culled as excess inventory, therefore, morbidity and malformation information are not available for these fish. If the culled eggs or fry were from crosses exhibiting high occurrences of malformations, morbidity and/or mortality, which would be the expectation for a hatchery which is attempting to select for the most fit and fastest growing offspring, this would tend to skew the population of the fish remaining in the facility after culling towards one with lower occurrence of these parameters. Thus, data collected on later life stages in this facility may be biased to some extent, with the bias potentially increasing with the age of the fish.

ABT provided a large amount of retrospective data on the entire 2004 breeding season's fish. That year, roughly 19,000 AquAdvantage and roughly 6,000 "wild-type" fry were grown. Pre-smolt size fish exhibited similar numbers of developmental irregularities (0.3% and 0.5% for AquAdvantage and comparator fish, respectively). Mortalities were higher among comparator fish in these early stages (8.7% for AquAdvantage Salmon versus 18.5% for comparator fish), and mortalities were similar among animals once they reached larger grow-out weights (3.5% for AquAdvantage Salmon versus 2.0% for comparator fish).

Conclusion: There are no consistent gross differences in mortality and morbidity between AquAdvantage Salmon and non-GE Atlantic salmon in either a small, controlled study or a large-scale historical retrospective data evaluation. There is some residual uncertainty due to differences in culling procedures employed at the PEI facility from those that may be employed at commercial grow-out facilities. The Durability Plan includes monitoring, data collection, and reporting of abnormalities observed under commercial production and grow-out conditions to address this residual uncertainty.

Table 5. Average (%) Survival to First-Feeding for ABT Salmon (TX x SC) & Non-GE (SC x SC) Salmon from the 2001-2006 Year-Classes*

Spawning Year	ABT Salmon Crosses			Non-GE Crosses	
	TX ♂ x SC ♀	SC ♂ x TX ♀	TX ♂ x SC ♀ PS (3n)	SC ♂ x SC ♀	SC ♂ x SC ♀ PS (3n)
	Average % survival & range of % survival *(Min-Max)* for number of crosses (n)				
2001	75 (37-95) n=14	nd	70 (37-94) n=13	67 (22-95) n=9	58 (20-92) n=4
2002	71 (60-81) n=8	nd	21 (10-38) n=6	72 (45-88) n=9	45 (10-87) n=5
2003	42 (2-82) n=29	nd	44 (2-86) n=25	46 (8-84) n=8	41 (18-88) n=9

Table 5. (Continued)

Spawning Year	ABT Salmon Crosses			Non-GE Crosses	
	TX ♂ x SC ♀	**SC ♂ x TX ♀**	**TX ♂ x SC ♀ PS (3n)**	**SC ♂ x SC ♀**	**SC ♂ x SC ♀ PS (3n)**
2004	54 (31-73) n=8	24 (4-50) n=3	50 (33-58) n=4	59 (13-89) n=7	57 (52-62) n=2
2005	48 (12-90) n=10	49 (26-64) n=8	37 (4-85) n=3	17 (6-31) n=5	nd
2006	70 (10-98) n=12	86 (53-97) n=4	95 (91-98) n=8	95 (94-96) n=4	94 (92-95) n=3

* Abbreviations: ***TX***, Treated (ABT Salmon); ***SC***, Sponsor Control (comparator non-GE); ***PS** (3n)*, pressure shocked (i.e., triploid); ***nd***, no crosses set up.

g. Clinical Pathology Assessments

We are primarily interested in hematology and serum chemistry comparisons affected by the GE status of the study population. With respect to the clinical pathology values obtained in the animal safety study, we found that all differences between AquAdvantage Salmon and non-GE salmon can be explained by triploidy, seasonality, growth conditions at the time of harvest, or a combination of these factors. In other words, no clinically relevant differences in the serum chemistry or hematology values for AquAdvantage Salmon as compared with contemporaneous non-GE Atlantic salmon are clearly attributable to the GE construct.

An additional overarching consideration in interpreting the clinical pathology data collected in this study is the well known effect of triploidy, including increasing cell size with resulting effects on other parameters (Benfey, 1999; Cal et al., 2005; Dorafshan et al., 2008). For example, erythrocyte counts are generally lower for triploid fish than for diploid fish, with corresponding decreases in packed cell volume (PCV), hematocrit, and hemoglobin widely reported in triploid fish relative to diploid comparators.

The available clinical pathology data for Atlantic salmon, while quite extensive relative to available data for other fish species, is relatively limited compared to similar data for terrestrial species. Serum chemistry and hematology are not assays routinely conducted by aquaculture facilities so the historical data from the literature for these clinical pathology assessments are primarily useful for understanding the breadth of the values considered "normal" under a variety of growth and aquaculture conditions. There are also notable gaps in published data and the range of clinical pathology values that have been reported.

Among the cited references in the animal safety study report, one is particularly relevant to this evaluation. Cogswell *et al.* (2001) have previously published the hematology values of diploid and triploid growth hormone (GH) GE Atlantic salmon produced by ABT. They report that triploid erythrocytes are significantly longer and proportionately thinner than diploid erythrocytes for both GE and non-GE genotypes. The authors speculated that GE fish may produce erythrocytes with higher surface area to volume ratio in response to their elevated metabolic rates. No other major hematological differences were observed between GE and non-GE salmon of the same ploidy.

In the safety study, samples obtained from each enrolled animal were analyzed for clinical chemistry and complete blood count parameters. These data included hemoglobin, hematocrit, platelet count, neutrophils, lymphocytes, monocytes, glucose, sodium, potassium, chloride, alanine aminotransferase, aspartate aminotransferase, total bilirubin, creatine kinase, total protein, albumin, globulin, albumin/globulin ratio, calcium, inorganic phosphorus, cholesterol, and osmolality (see Phenotypic Characterization Appendix, below).

Overall, the range of values for the various parameters was comparable for the AquAdvantage and non-GE salmon in the study. In several cases, however, as illustrated by the figures in the Phenotypic Characterization Appendix, below, there appeared to be differences in the values for specific subgroups of the fish. Often these differences were identified by statistical analyses when focusing on the comparison between the GE triploids and the non-GE diploids. Several analytes were identified as exhibiting a statistically significant difference ($p < 0.05$) when evaluating the effect of the AquAdvantage construct; other statistically significant differences were also identified, but were not related strictly to the GE status of the fish. For example, the hematocrit and platelet count were at higher levels in at least some of the diploids as compared with triploids. This effect of triploidy is expected and is well documented in the literature as discussed above. The other differences are discussed below.

Analysis of the hematology values showed a difference in the relative level of lymphocytes and neutrophils for the diploid AquAdvantage Salmon as compared to all of the other market size salmon in this study. However, the relative levels of these white blood cells in the market size diploid AquAdvantage Salmon were within the range of levels reported in the literature and are comparable to the levels for satellite controls sampled at the same time. We conclude that these differences in hematology reflect differences in growth conditions at the time of sampling and are not attributable to the GE status of the fish.

Similarly, initial comparisons of the protein values (albumin, globulin, total protein and albumin:globulin ratio) for the market size fish identified a statistically significant difference between the GE and non-GE fish. These differences between the groups were small, and unlikely to be biologically relevant. Furthermore, the values determined for the market size AquAdvantage Salmon were comparable to published values as well as the values determined for the age matched satellite control non-GE fish. Similar apparently statistically significant differences were initially identified for calcium, cholesterol, phosphorous, and total bilirubin, but were not reflective of a difference attributable to the test article when the age-matched comparators and historical data are considered.

Aspartate aminotransferase (AST) was identified as being statistically significantly different when non-GE fish were compared with the AquAdvantage Salmon; it is however, lower in the AquAdvantage Salmon than in the non-GE comparators. Elevation of AST is often used clinically as an indicator of tissue damage, so this difference is not likely to be indicative of a health problem in the AquAdvantage salmon. Further examination of the AST values in the comparators indicated that this elevation in AST levels for the non-GE group is likely influenced by one of the market sized non-GE fish, and thus is not attributable to the GE construct.

Finally, there was a statistically significant difference between the glucose level of the ABT salmon and non-GE market size comparator fish. Examination of the graphically presented data showed that the overall values were not grossly different for the groups, but that the glucose values for the AquAdvantage Salmon tend to be a bit lower than for the non-GE comparators. The glucose values reported here were lower than the values reported in the literature. This could reflect a difference in handling of the fish (fasting, sedation, tank conditions), or the samples (glucose levels are typically reduced in samples that are not processed immediately as the blood cells in the samples will metabolize the glucose). Some of the values were so low for all of the market size fish groups (<40 mg/dl) that an artifactual source such as a longer holding time seems likely. Values this low, if real, would likely result in observable behavior deficiencies (swimming, feeding, etc), which were not noted, or compromised growth which was not the case. Alternatively, reports in the published literature (see discussion below in Part B.1.k). Other Phenotypic Characteristics) indicate that the GE Atlantic salmon have a higher metabolic rate than non-GE comparators, which would be consistent with a lower serum glucose level – or increased utilization of glucose in samples held before processing. In conclusion, the statistical difference in reported glucose values is unlikely to represent a biologically relevant finding.

Conclusion: We found no clinically relevant differences in the serum chemistry or hematology values for AquAdvantage Salmon compared with contemporaneous non-GE Atlantic salmon that are clearly attributable to the GE construct.

h. Macroscopic and Microscopic Evaluation (Other Than Gross Morphology)

Macroscopic and microscopic observations (other than gross morphology of the musculoskeletal system as described above) of the EO-1α lineage of Atlantic salmon were provided by ABT.

Observations included the 2001-2005 year classes, and were collected as part of routine health evaluations of the broodstock development program. As fish were found dead, moribund, or culled, selected individuals were subject to necropsy and diagnostic histopathology and bacteriology as deemed necessary by ABT and consulting fish pathologists at an independent veterinary pathology laboratory[8].

Among the necropsy and histopathologic findings, spontaneous skeletal deformities were noted, including dorsoventral and lateral deviations of the vertebral column. In addition, malformations of the head, primarily lower jaw, were observed. These macroscopic observations have been described above in this evaluation by year class and include GE and non-GE diploid and GE and non-GE triploid fish. Information from 2006 and 2007 year classes has been summarized together with information from 2003-2005 year classes with respect to the rate of irregular external appearance in Table 4.

Microscopic observations from earlier generations of diploid and triploid AquAdvantage Salmon and non-GE comparators document a variety of inflammatory and degenerative lesions that are mostly consistent with diseases of intensively-reared fish. The data and information provided were assembled from fish production records and reflect a variety of crosses and husbandry conditions over several years. The range and severity of histopathologic lesions, morphologic diagnoses, and etiologic diagnoses do not appear to indicate a difference in frequency between GE and non-GE fish. Although this is not an adequate and well controlled study due to the variability of husbandry conditions, numbers of fish crosses, and long time course, this information is nonetheless considered as part of our weight-of-evidence evaluation, and contributes to our understanding of the effect of the AquAdvantage construct on the fish. In addition, significant morbidity and mortality could be masked as a result of the rigorous culling practices necessitated by the size of the broodstock facility. On the other hand, the information provided encompasses a large number of animals over many generations and year classes, so there is a level of inferential value for this information to the general situation of rearing of Atlantic salmon containing the

AquAdvantage construct. These observations are generally consistent with those of the animal safety study. Observations for fish in the 2007 year class were collected as part of the animal safety study. The experimental design is described above. Table 6 summarizes significant lesions.

Macroscopic (Gross) Observations

Gill Abnormalities

Gill filaments (primary lamellae) were truncated (shortened, incompletely developed) or absent; deformities of the gill arch itself, with or without scarring; and gill thickening or opacity were noted by ABT. Structural gill arch abnormalities (truncated or absent filaments and gill arch deformities) were most often correlated microscopically with truncated or absent filaments, whereas gill arch thickening was correlated with increased segmental hyperplasia of the lamellar (secondary lamellae) epithelium according to ABT. Gill abnormalities were substantially more prevalent among triploids than diploids. There is no observed effect on gill structure that is attributable to the GE status of the fish in the study.

Fin Abnormalities

Erosions, shortening, twisting (torsion), nodules and bifurcation were noted by ABT. These lesions were distributed among a variety of fin types (pectoral, dorsal, pelvic, caudal) among different fish. Except for a fin nodule that was correlated with an epidermal cyst (Male, Diploid, SP-CON), and a shortened, twisted dorsal fin that was correlated with a skeletal deformity (Female, Triploid, SP-CON), the study pathologist was unable to associate specific microscopic changes with macroscopic fin lesions (in four instances, fin tissues with gross findings were not available for microscopic examination). Triploids exhibited a higher prevalence than diploid for fin abnormalities. There is no observed effect on fin structure that is attributable to the GE status of the fish in the study.

Heart Abnormality

Loss of pyramidal profile (cardiac shape abnormalities) of the heart was noted in some fish by ABT. The prevalence was higher among triploids according to ABT. No microscopic correlates were observed. There is no observed effect on heart structure that is attributable to the GE status of the fish in study.

Table 6. Prevalence of Selected Gross Observations

	Diploid						**Triploid**					
Group	**SP-CON**		**SAT-CON**		**ABT Salmon**		**SP-CON**		**SAT-CON**		**ABT Salmon**	
Sex	**M**	**F**	**M**	**F**	**M**	**F**	**M**	**F**	**M**	**F**	**M**	**F**
Number Examined	6	6	3	3	6	6	6	6	3	3	6	6
Gill Arch Structural Abnormalities	2	2	-	-	-	-	5	6	2	-	5	5
Gill Arch Thickening/ Opacity	2	2	-	-	-	-	3	5	1	1	1	2
Fin Abnormalities	1	-	-	-	-	-	-	3	-	1	2	1
Heart Abnormalities	-	-	1	-	-	-	-	1	1	1	-	1
Jaw Erosions	-	-	-	-	3	1	-	-	-	-	-	-

SP-CON = Sponsor control, SAT-CON = Satellite control

Jaw Erosions

Jaw erosions were noted by ABT exclusively in male (three of six) and female (one of six) diploid AquAdvantage Salmon. No jaw erosions were noted among triploid fish irrespective of GE status.

Microscopic Observations

Focal Inflammation

Foci of inflammation, which were generally minimal to mild, were observed in a variety of tissue types by ABT. Inflammation was most frequently characterized as granulomatous, consisting of chiefly macrophages in spherical nodular aggregates, with or without multinucleated giant cells or central areas of necrosis. Other types of inflammatory lesions (acute, chronic active, necrogranulomatous, pyogranulomatous) were less regularly observed. The most commonly affected sites for inflammation were the abdominal mesentery, cranium, and trunk kidney, see Table 7. Etiologic agents were not evident in any of the lesions. According to ABT, the prevalence of focal inflammation was higher among diploid fish than triploid and higher among diploid (and to a lesser extent triploid) AquAdvantage Salmon compared with either size-matched (SP-CON) or age-matched (SAT-CON) controls.

Gill Lesions

ABT reported a higher prevalence of gill lesions among triploid fish than diploid fish. The lesions included structural abnormalities of the gill filaments along with increased segmental lamellar epithelial hyperplasia. The structural abnormalities were truncated or absent gill filaments. In some instances the abnormal filaments also demonstrated inflammation. Absent gill filaments occurred most often at the apex of the gill arch. Little, if any, inflammation of the gill arch was noted.

Information on the gill morphometry of GH transgenic Atlantic salmon has been reported by Stevens and Sutterlin (1999). Although these salmon were GH transgenic Atlantic salmon produced by ABT, from the information provided, it is impossible to determine whether or not they were in fact early generations of what is currently known as AquAdvantage Salmon. The authors found that many of the morphological features of the respiratory system of these salmon are larger than those of similarly-sized comparator salmon. For example, the gill surface area available for respiratory exchange in the transgenic salmon was found to be about 1.24 times that of comparator salmon. The increase in gill exchange area was due largely to a relatively uniform increase in the length of each gill filament. However, the authors reported that there were no obvious differences between the two groups of salmon in gill morphology when viewed with a dissecting microscope.

Ectopic Mineralization

There was substantially higher prevalence of soft tissue mineralization affecting multiple tissue types in SAT-CON triploids and AquAdvantage triploids compared to SP-CON triploid and diploids. Among affected fish, females generally had a higher prevalence of mineralization than males. The most commonly affected sites were the eye, heart, liver, and trunk kidney. Most instances were graded as minimal, although some lesions were mild to moderate. A few mineralized lesions of the urinary tract were noted grossly at necropsy.

Table 7. Prevalence of Inflammation in Various Tissue Types[a]

	Diploid						Triploid					
Group	**SP-CON**		**SAT-CON**		**ABT Salmon**		**SP-CON**		**SAT-CON**		**ABT Salmon**	
Sex	**M**	**F**	**M**	**F**	**M**	**F**	**M**	**F**	**M**	**F**	**M**	**F**
Number Examined	6	6	3	3	6	6	6	6	3	3	6	6
Body wall	1	1	-	-	-	-	-	-	-	-	-	-
Cranium	-	3	1	-	4	1	-	-	-	-	-	1
Distal Intestine	-	-	-	1	-	-	-	-	-	-	-	-
Eye	1	-	-	-	1	1	-	-	-	-	-	-
Gall Bladder	-	1	-	-	-	-	-	1	-	-	-	-
Head Kidney	-	-	1	-	-	-	1	-	-	-	1	-
Heart	-	-	-	-	-	2	-	1	-	-	1	-
Liver	1	2	-	-	1	1	1	1	1	-	-	1
Spleen	-	-	1	1	1	-	1	2	2	-	1	-
Stomach	1	-	-	-	-	-	-	-	-	-	-	-
Swim Bladder	-	-	-	1	1	1	1	1	-	-	-	-
Testes	-	na	-	na	-	na	-	na	1	na	-	na
Trunk Kidney	1	1	2	-	2	4	1	1	1	1	3	1
Number Examined	1	1	0	0	3	2	1	0	1	1	3	1
Abdominal Mesentery	1	1	-	-	3	2	1	-	1	1	3	-
Number Examined	1	1	1	0	1	1	1	0	0	0	2	0
Pancreas	1	1	1	-	1	1	1	-	-	-	2	-
Number Examined	0	0	0	0	1	0	0	0	0	0	0	0
Urinary Bladder	-	-	-	-	1	-	-	-	-	-	-	-

[a] Based on number of animal affected per group; na= Not applicable
SP-CON SAT= Sponsor control; -CON = Satellite control

Hepatocellular Vacuolization

Hepatocellular vacuolization ranged from minimal to moderate and tended to be higher in triploid salmon as compared to diploids. Vacuolization was characterized by single or multiple, variably-sized, discrete, round, sharply-defined spaces within the hepatocyte cytoplasm. Larger vacuoles displaced the nucleus toward the periphery of the cell.

Discussion of ABT's Results in the Animal Safety Study

The experimental design of this study adequately addresses the situation for age-matched vs. size-matched comparators; however, the effects of seasonality and small sample size limits the inferential value of the conclusions that can be drawn from this study. Effects of seasonality may be evident in the clinical pathology data evaluated above. The effect of seasonality has been suggested as a possible explanation for differences in the leukocyte profile, specifically lymphocytes and neutrophils, particularly among AquAdvantage and SAT-CON diploids. The effect of small sample size has been suggested by ABT as a limitation on the interpretation of jaw erosions.

Of the macroscopic and microscopic lesions observed, most appear to be associated with the induction of triploidy. Morphologic abnormalities of the axial skeleton, fins, opercula, and gills have been documented among a variety of finfish in the literature and in triploid Atlantic salmon in particular. Although gill abnormalities, often accompanied by skeletal, jaw, and opercular malformations, have been most commonly reported in the literature for triploid Atlantic salmon, the prevalence of fin deformities was increased in this study.

Microscopic lesions of the gills are extremely well correlated to gross observations and are consistent with those described in the literature for triploid fish. Ectopic mineralization, seen with higher prevalence among triploids in this study, has not previously been documented in the literature as being an effect of this process. Hepatocellular vacuolation in many species, including fish, is a reflection of lipid metabolism within the body. ABT notes its presence here with higher prevalence among triploids, in addition the literature notes some cases of small livers and lower liver RNA among triploids. As ABT points out, there are many factors that may influence the deposition and mobilization of lipid stores that might ultimately result in hepatocellular vacuolation.

An increased prevalence of focal inflammation in various tissue types in AquAdvantage Salmon has the strongest correlation with the presence of the AquAdvantage construct among the findings in this study. That these fish may have been immunocompromised as a result of seasonality or other factors confounds the interpretation of these findings.

Conclusions: Macroscopic observations of gill, fin, and heart abnormalities are most likely due to the induction of triploidy, rather than as a result of fish containing the AquAdvantage construct. The most likely cause of the jaw erosions, observed exclusively in male and female diploid AquAdvantage Salmon, is the presence of the AquAdvantage construct.

Microscopic observations of gill lesions and ectopic mineralization are most likely associated with the induction of triploidy. The increased prevalence of focal inflammation is most likely due to the presence of the AquAdvantage construct.

Although the presence of the AquAdvantage construct appears to have increased the prevalence of jaw erosions and focal inflammation in adult fish, we concur with ABT's assessment that these findings are of low magnitude and not likely to be debilitating to fish in a production setting.

i. Disease Resistance

ABT briefly describes the available information on the disease resistance of AquAdvantage Salmon. Comprehensive disease challenge studies have not been conducted on these fish; however, a limited study on 20g fish was performed to determine if the presence of the AquAdvantage construct alters the disease resistance of the AquAdvantage salmon to furunculosis (*Aeromonas salmonicida*) compared to size matched non-GE salmon. In this study, there was an earlier peak in the mortality of the AquAdvantage Salmon following challenge (days 12-15) relative to the comparators (days 14-21), but overall there was no obvious difference in mortality profiles between the two according to the study investigators.

Conclusion: The limited available information does not indicate a significant change in disease resistance of AquAdvantage Salmon relative to non-GE comparators.

j. Smoltification and Seawater Survival

AquAdvantage Salmon are not currently proposed for grow-out in seawater, but there are currently data available from studies under contained conditions to address the ability of these fish to undergo successful smoltification and survive a transfer from freshwater to seawater. This information is useful for environmental assessment purposes to evaluate survival under various escape scenarios and is important if seawater grow-out is proposed in the future. Pilot studies suggest that diploid AquAdvantage Salmon undergo a normal smoltification process and will survive if transferred from freshwater to seawater (see Table 8 below). According to the report, survival of diploids in the body-weight range examined was consistent with commercial experience with non-GE Atlantic salmon. Survival was very high ($\geq$ 98%) when the fish weight at transfer was at least 150 grams. Comparable data for triploid AquAdvantage Salmon were not available, but there have been reports in the literature that the survival rate of triploids in saltwater is lower than that of diploids (Benfey, 2001; Galbreath and Thorgaard, 1995; O'Flynn et al., 1997).

Additional data on smolt development in Atlantic salmon genetically engineered with growth hormone have been published by Saunders *et al.* (1998); these findings are consistent with the data described above for AquAdvantage Salmon. GE individuals that approached smolt size (16 cm) were able to survive for greater than 96 hours following direct transfer from freshwater to full strength seawater with a salinity of 35%, while their normal, non-GE siblings were smaller (<10 cm) and survived less than 24 hours following a similar transfer to seawater. In addition, GE salmon exposed to various temperature-photoperiod conditions were able to complete the smoltification process under conditions that would inhibit or delay completion of smolting in non-GE Atlantic salmon. After transfer to seawater, GE salmon exhibited satisfactory survival and growth for an additional 4 months (when observations were terminated).

Conclusion: The limited available information suggests that diploid ABT salmon of smolt size will survive and grow normally following transfer from freshwater to seawater, indicating that basic aspects of the physiology of the GE salmon have not been altered and that the presence of seawater would not act as physical barrier to survival and establishment. Information on smoltification for triploid GE salmon is currently lacking.

k. Other Phenotypic Characteristics

A wide variety of additional phenotypic characteristics of Atlantic salmon genetically engineered with growth hormone (GH) (GH transgenic Atlantic salmon)[9, 10] have been studied and reported in the scientific literature by investigators at ABT and academic research institutions in Canada. These characteristics include feed consumption, foraging and predator avoidance (Abrahams and Sutterlin, 1999), gill morphology (Stevens and Sutterlin, 1999), gut morphology (Stevens et al., 1999), myogenesis and muscle metabolism (Levesque et al., 2008), metabolic rate (Cook et al., 2000b), respiratory metabolism and swimming performance (Stevens et al., 1998). In general, these studies found that pre-smolt GH transgenic Atlantic salmon had higher rates of myogenesis, muscle metabolism, and oxygen consumption than non-GE comparators, as well as altered morphology of some body structures. These findings are not inconsistent with the rapid growth phenotype. Because these data are limited to scientific literature investigating GH transgenic Atlantic salmon, but not necessarily the specific fish considered for approval in this application, these data are given less weight than the controlled studies presented above. In many cases, these data are the only data available addressing these phenotypic characteristics.

Stevens et al. (1998) found that, pre-smolt GH transgenic Atlantic salmon have a higher oxygen uptake during routine culture conditions and during forced swimming activity relative to similar sized comparators. Overall, the oxygen uptake of GE fish was 1.7 times that of comparators over the course of a day. These fish also had a higher critical oxygen concentration. (Critical oxygen concentration is the concentration in water at which oxygen uptake by fish becomes limited by the oxygen supply (i.e., the concentration threshold where the oxygen uptake rate starts to decrease)). The critical concentration for GE fish was 6 mg/L vs. 4 mg/L in comparator fish. This higher critical oxygen level for GE fish has potential implications for growers and for the establishment and survival of fish if they somehow escape from grow-out facilities.

Particularly in areas where water temperatures are elevated[11], water oxygen levels may be below the critical level; any escaped fish will likely be adversely affected and may not survive for extended periods of time. In addition, Stevens et al. (1998) caution that future growers of growth enhanced salmon should be prepared to either deliver more water or more oxygen in the water per unit of biomass of GE fish compared to that required by non-GE salmon. Based on the information reviewed, AquAdvantage Salmon may have reduced tolerance for low dissolved oxygen content.

Table 8. Survival of AquAdvantage Fry-Smolt Following Transfer to Seawater

Year Class	Transfer Date	Transfer Weight	Survival
1999	Nov 2000	30 g	85%
2001	Dec 2002	57 g	80%
2002	Oct 2003	161 g	98%
2005	Jul 2006	150 g	99%

GE Atlantic salmon have been found to have much higher rates of feed consumption than comparators, and were more willing to feed in the presence of a predator (Abrahams and Sutterlin, 1999). In terms of gut morphology, one group has reported that GH transgenic Atlantic salmon have more (and longer) intestinal folds and a larger digestive surface area than size-matched non-GE comparators (Stevens et al., 1999). Most morphological features of the intestine and of the pyloric caeca of GE salmon were larger than those of comparator salmon. However, the animal safety study specifically addresses the observations of Stevens *et al.* and did not report any significant macroscopic or microscopic differences between AquAdvantage Salmon and comparator (non-GE) with respect to gut morphology.

Growth rates, body composition, and feed digestibility/conversion efficiency have been studied in pre-smolt (8 - 55 g) GH transgenic Atlantic salmon by investigators from ABT and the Atlantic Veterinary College (Cook et al., 2000a). In this study, GE fish exhibited a 2.62 to 2.85-fold greater rate of growth compared to non-GE fish over the body weight ranges examined. In addition, gross feed conversion efficiency in pre-smolts was improved by approximately 10% relative to non-GE comparator fish. Body protein, dry matter, ash, lipid and energy were significantly lower in the GE salmon pre-smolts relative to comparators, while moisture content was significantly higher. Compositional analysis of AquAdvantage

Salmon will be specifically addressed during the food safety evaluation (see Section VII below).

A comprehensive, comparative examination of the cardiorespiratory physiology of post-smolt from the fifth generation of GH transgenic salmon has also been conducted by Deitch *et al.* (2006). In this study, GH transgenic salmon had an 18% lower metabolic scope, 25% higher standard oxygen consumption, and 9% reduction in critical swimming speed relative to size-matched non-GE comparators. This decreased metabolic capacity/performance occurred despite the fact that GH transgenic salmon had a 29% larger heart and increased cardiac output. Because gill surface area was the only cardiorespiratory parameter that was not enhanced in these salmon, it was suggested that gill oxygen transfer may have been limiting.

Conclusion: A number of phenotypic characteristics are altered in GH transgenic salmon relative to non-GE comparators. Many of these changes, for example increased growth rate, are the intended and expected effects of introduction of this GH construct into the salmon and so are desirable. None of these changes, as described in GE Atlantic salmon, would be expected to adversely affect the animal health or safety of AquAdvantage Salmon under normal conditions of commercial grow-out if adequate water oxygen levels are maintained. Some of the reported changes would potentially make these fish less fit and less likely to survive if they were to escape from grow-out facilities.

l. Monosex (All-Female) Population

Although the product definition for AquAdvantage Salmon describes the fish intended for marketing to be monosex (all-female), the studies that have been conducted to date have included mixed populations of both males and females. In most of the early studies, no attempt was made to determine the gender of the fish. In the large animal safety study, where fish were identified by gender, it was possible to evaluate the effects of the rDNA construct by sex. No significant differences in gross morphologic or microscopic lesions were apparent in that study.

In order to ensure an all-female population, the sponsor has recently utilized a gynogenesis method. The adequacy of the gynogenesis process and overall production plan is reviewed as a part of the Durability Plan assessment. Based on the information provided by the sponsor and considering the physiologic mechanisms of the gynogenesis process, we believe the phenotypic characteristics of a mixed gender population adequately represent the range of phenotypic characteristics expected in a monosex (all female) population.

C. Addressing the Risk Questions

Risk Question 1: Is There Direct or Indirect Toxicity to the Animal?

There is evidence of minimal direct effects in fish containing the AquAdvantage construct in the form of an increased frequency of skeletal malformations, and increased prevalence of jaw erosions and multisystemic, focal inflammation. There is also evidence from the scientific literature of minimal indirect effects in the form of increased cardiac output and reduced tolerance to low dissolved oxygen (DO) concentrations. These effects of the AquAdvantage construct are likely to impact the overall fitness of AquAdvantage Salmon in the natural environment. The consequences of these effects to AquAdvantage Salmon in a production setting are likely to be small and within the range of abnormalities affecting rapid growth phenotypes of Atlantic salmon.

We further note that although a lack of information contributes to uncertainty regarding the rate of abnormalities in commercial grow-out facilities, given restrictions regarding the number of animals that may be raised under the investigational phase, there is no practical way ABT could have generated the appropriate data without producing – and destroying – commercial lots of fish. Nonetheless, we believe that incorporating an appropriate surveillance/durability plan will provide sufficient data and information to the Agency to minimize this uncertainty.

Conclusion: The phenotypic characterization of AquAdvantage Salmon has been adequately addressed according the risk-based, hierarchical system established for the evaluation of GE animals. There is sufficient information to support the safety of the construct to the AquAdvantage Salmon. No significant adverse outcomes were noted as the result of the incorporation of the AquAdvantage construct; therefore we conclude that no significant hazards or risks have been identified with respect to the phenotype of the AquAdvantage Salmon.

Risk Question 2: Are There Phenotypic Characteristics That Provide Hazard Identification for Other Steps in the Evaluation?

No hazards have been identified specifically for the genotypic and phenotypic durability, environmental and food/feed safety, or claim validation evaluations. With respect to the environmental safety evaluation, several phenotypic changes have been identified that are consistent with the presence of the AquAdvantage construct and appear to result in decreased fitness (e.g., increased oxygen requirements, decreased critical swimming speed, lower metabolic scope, etc.). These changes are expected to impact survival and establishment should any AquAdvantage Salmon escape from commercial production facilities.

Conclusion: No phenotypic characteristics have been identified that would provide hazard identification for other steps in the evaluation based on the current product definition and development plan. We note that alterations in some of the phenotypic characteristics reported in the scientific literature for GH transgenic salmon may alter fitness and should be considered in the Environmental Safety evaluation (see Section VIII, below).

Risk Question 3: What Are the Risks to the User (User Safety)?

No data provided in the file suggests that there are any additional risks to handler safety above that of commercially farmed Atlantic salmon.

Conclusion: There is no risk to user/handler safety associated with AquAdvantage Salmon.

Risk Question 4: What Are the Risks from Any Components of Any Biological Containment Strategy?

Induction of triploidy in AquAdvantage Salmon contains increased risk of gill, fin, and heart abnormalities and ectopic mineralization. The severity of these effects is generally minimal and is not expected to have a significant consequence in a production setting. A reduction in growth characteristics has often been reported in the literature associated with the induction of triploidy in salmonids. The increased growth rate of the AquAdvantage phenotype may mitigate some effects of the triploidy procedure.

As discussed above, the effects of triploidy on AquAdvantage Salmon are no different than that observed with comparator salmon. Triploidy is a common aquaculture technique regularly used in practice.

Conclusion: There are no risks to AquAdvantage Salmon from triploidy that are not already present in triploid-based aquaculture systems.

D. Conclusion

We have conducted a weight-of-evidence evaluation of the AquAdvantage Salmon phenotype drawing on data from a number of sources. This evaluation has used four sources of data and information. As discussed in the weight-of-evidence section, we have placed the most emphasis on controlled studies conducted on the specific animals being considered for approval. We have also considered other non-controlled studies on these same animals, as well as historical hatchery records and data for these animals. Finally, we have evaluated studies reported in the scientific literature investigating these same animals or their relatives.

Our final conclusions are presented immediately below, and followed by summary justifications:

- The phenotypic characterization of AquAdvantage Salmon has been adequately addressed according the risk-based, hierarchical system established for the evaluation of GE animals.
- There is sufficient information to support the safety of the construct to the AquAdvantage Salmon.
- No unique adverse outcomes were noted as the result of the incorporation of the AquAdvantage construct; therefore we conclude that no significant hazards or risks have been identified with respect to the phenotype of the AquAdvantage Salmon.
- There is some uncertainty associated with the rate of abnormalities that may occur in commercial grow-out facilities due to data derived from the constrained conditions of broodstock facility. Increased mild abnormalities that have been noted are likely within the range observed in rapid growth phenotypes of non-genetically engineered Atlantic salmon.
- Because of increased metabolic demands of the rapid growth phenotype, it may be useful to provide labeling recommendations for ensuring that the animals are fed to satiety and that dissolved oxygen is carefully monitored to provide optimal growth conditions.
- Any adverse outcomes that have been noted (e.g., jaw malformations, increased metabolic demand) will likely render AquAdvantage Salmon less fit in a competitive environment. These findings should be address in the environmental assessment.

- Based on the information in this evaluation and considering the physiologic mechanisms of the gynogenesis process, we believe the phenotypic characteristics of a mixed gender population adequately represent the range of phenotypic characteristics expected in a monosex (all female) population.

No significant hazards have been identified and the phenotype is stable over at least six generations. Residual uncertainty regarding rates of abnormalities in commercial grow-out facilities suggests a safety surveillance program be included in the durability plan for AquAdvantage Salmon. Although no specific hazards have been identified, we note that alterations in some of the phenotypic characteristics reported in the scientific literature for GH transgenic salmon may alter fitness and should be considered in the Environmental Safety evaluation (see Section VIII below).

Post-approval monitoring and reporting as part of the Durability Plan is recommended to address uncertainties associated with the incidence of malformation rates in the early life-stages of AquAdvantage Salmon under commercial grow-out conditions. This monitoring should be conducted with several year-classes of salmon at the ABT grow-out facility in Panama. Additional data on malformations, morbidity and mortality in culled fish at the PEI broodstock facility would also decrease uncertainties.

VI. Genotypic and Phenotypic Durability

A. Overview

This section evaluates the genotypic and phenotypic durability of GE Atlantic salmon containing the AquAdvantage construct at the α- locus as well as the plan to ensure that GE animals in commerce post-approval remain equivalent to those evaluated for safety and effectiveness prior to approval.

In general, FDA evaluates new drugs for safety and effectiveness once: prior to approval and marketing. It is presumed that if the post-approval regulated article is equivalent to the article evaluated for safety and effectiveness during the pre-approval evaluation, then the post- approval article will have the same safety and effectiveness characteristics. Therefore, it is incumbent upon the sponsor to adequately demonstrate their ability to provide a consistent product and to put into place adequate procedures and testing to ensure that future products are, in fact, equivalent to those evaluated for safety and effectiveness. Additionally, in the event that the sponsor makes changes to the application after approval, they must submit those proposed changes in a supplemental application.

There are three risk questions posed during this step:

i. Is the genotype changing over the lifespan of the animal or product such that it would affect the risks associated with the product?
ii. Is the phenotype changing over the lifespan of the animal or product such that it would affect the risks associated with the product?

iii. Is there a plan in place to ensure that over time the phenotype and genotype will not change, or if it does, are there procedures in place to provide for either a remedy or a risk assessment of the new animal?

The review process described in Guidance 187, *Regulation of Genetically Engineered Animals Containing Heritable rDNA Constructs*, assumes that the initial levels of review primarily identify and characterize potential hazards associated with the GE animal and that successive levels of review consider any hazards that have been previously identified. Evaluations at the earlier levels did not identify hazards that impact the durability assessment or plan. Therefore no risk questions beyond that stated above need to be addressed in this section. The Phenotypic Characterization step (see Section V above) identified some areas of uncertainty, and recommendations involved addressing these uncertainties through a post-approval surveillance program at the Panamanian facility. We incorporated this program as a part of the durability plan for AquAdvantage Salmon as discussed below.

After evaluating data submitted by ABT, we conclude that the data submitted support the Genotypic and Phenotypic Durability portion of the hierarchical review of AquAdvantage Salmon, and that the proposed Durability Plan is acceptable.

B. Evaluation

1. Is the Genotype Changing over the Lifespan of the Animal or Product Such That It Would Affect the Risks Associated with the Product?

The genotypic durability assessment was conducted during our evaluation of the Molecular Characterization of the GE Animal Lineage (see Section IV above). We found that the genotype was durable over seven generations.

Conclusion: The genotype is not changing between generations and is durable. The genotype is not changing in any manner that would impact the other risk questions in this or other steps of the hierarchical review process for AquAdvantage Salmon.

2. Is the Phenotype Changing over the lifespan of the Animal or Product Such That It Would Affect the Risks Associated with the Product?

The phenotypic durability assessment was conducted during the Phenotypic Characterization step of the hierarchical review process (see Section V above). We found that the phenotype was durable over six generations.

Conclusion: The phenotype is not changing between generations in a manner that impacts the risk questions for this or other steps of the hierarchical review process for AquAdvantage Salmon. Therefore, the phenotype of the AquAdvantage Salmon is durable.

3. Is There a Plan in Place to Ensure That over Time the Phenotype and Genotype Will Not Change, or if It Does, Are There Procedures in Place to Provide for Either a Remedy or a Risk Assessment of the New Animal?

A durability plan consists of several components. First, a plan should contain a listing of one or more characteristics critical to the durability of the final product as well as methods, testing schedules, and specifications for each of these characteristics. Second, the plan should discuss procedures that will be carried out in the event that an individual test result does not meet its specification. Third, the plan should include a commitment from the sponsor to withhold or withdraw from the market any product that does not meet all of the durability specifications. Finally, additional tests and monitoring procedures may be included as a part of the durability plan in order to address uncertainties from other steps of the product's evaluation.

In this section, we first present the proposed characteristics to be evaluated as a part of the durability plan. Then we discuss the overall testing schedule and production plan, followed by a detailed evaluation of each testing method and its validity. Procedures for out-of-specification results and ABT's withdrawal commitment are then evaluated. Finally, evaluation of an additional safety surveillance program is presented.

Proposed tests, schedules, methods, and specifications are described in Table 9.

a. Testing Schedule and Production Plan for Broodstock

ABT proposed a two-phased testing schedule for qualifying broodstock into their production program based on their overall plan for the production of AquAdvantage Salmon (the "production plan"). The first phase covers the three production cycles after approval. During this phase, all broodstock and the homozygous females used to generate the broodstock will be individually qualified. In the second phase (after the third production cycle), assuming no out-of-specification results during phase one, the testing schedule will be reduced to include only the homozygous females used to generate future broodstock. In the unexpected event that there is an out-of-specification result during phase one, the duration of that phase will be reconsidered.

ABT proposes to follow a specific production plan. The overall plan is presented in Figure 3.

Females homozygous for the AquAdvantage construct are used as the broodstock source for AquAdvantage Salmon. These fish are pre-qualified via confirmation of construct presence and construct stability via multiplex qualitative PCR and for construct copy number using Southern blot. Eggs from qualified homozygous females are then subjected to gynogenesis, resulting in an all-female gynogen offspring population.

Gynogenesis is a procedure commonly used in modern salmonid aquaculture. Briefly, irradiated sperm are introduced to eggs, followed by a pressure treatment to result in diploid "twin" offspring. In this case, ABT uses Arctic char milt that has been irradiated so that no Arctic char DNA is present in the gynogen population. In the event that the milt irradiation was not successful, the offspring would be an Arctic char / Atlantic salmon hybrid. These fish are readily identifiable by their differential markings and phenotypic appearance. As such, these hybrid fish can be easily removed without extensive testing. Therefore, only offspring that appear to be Atlantic salmon will be gynogens of the homozygous females; these animals will be used in subsequent steps in the production plan. The all-female gynogen population is then masculinized through the application of 17-methyltestosterone, so that they produce milt instead of viable eggs.

We have considered the use of 17-methyltestoterone in the production of AquAdvantage Salmon. This use is fairly common in modern aquaculture to produce "neomales" homozygous for the construct. These neomales are then used to produce all female hemizygous population of GE fish. Because these steps are removed from the production of food, we do not consider the use of this compound to pose a food consumption risk to the human consumer of the marketed AquAdvantage Salmon product.

During the first three production cycles after approval, the resultant homozygous neomale broodstock fish will be subjected to confirmatory testing for construct presence, stability, and copy number. These homozygous, neomale gynogens comprise the commercial broodstock for AquAdvantage Salmon.

Table 9. Summary of Durability Plan

Characteristic	Sample Matrix	Method	Testing Parameters
Presence of AquAdvantage Construct	Blood	PCR	(1) Individual testing of homozygous females used for generation of broodstock, and (2) Each broodstock fish during the first three production cycles after approval
AquAdvantage Construct Stability	Blood	PCR	(1) Individual testing of homozygous females used for generation of broodstock, and (2) Each broodstock fish during the first three production cycles after approval
Copy Number	Blood	Southern Blot	(1) Individual testing of homozygous females used for generation of broodstock, and (2) Each broodstock fish during the first three production cycles after approval
Triploidy	Eyed-eggs	FACS	Composite sampling from multiple egg batches in a single 23L upwelling chamber; the number of triploid eggs in each lot must be sufficient to provide a high expectation that the lot consists of at least 95% triploid eggs.

Broodstock salmon, upon sexual maturity, are then out-crossed with non-GE Atlantic salmon females. Collection of the milt from neomale broodstock is a terminal procedure, as these fish lack seminal vesicles for normal spawning. Each neomale, therefore, may only be used for one production cycle. According to statements by ABT, milt from a single fish can fertilize approximately 50,000 eggs.

Fertilized eggs are then subjected to pressure shock treatment in order to render the offspring triploid, with two copies of the genome coming from the non-GE female and one copy from the homozygous neomale gynogen. Female triploid salmon are effectively reproductively incompetent, providing additional environmental and intellectual property safeguards. Samples from individual triploid batches are combined to assess the rate of triploid induction via fluorescence-activated cell sorting (FACS) analysis.

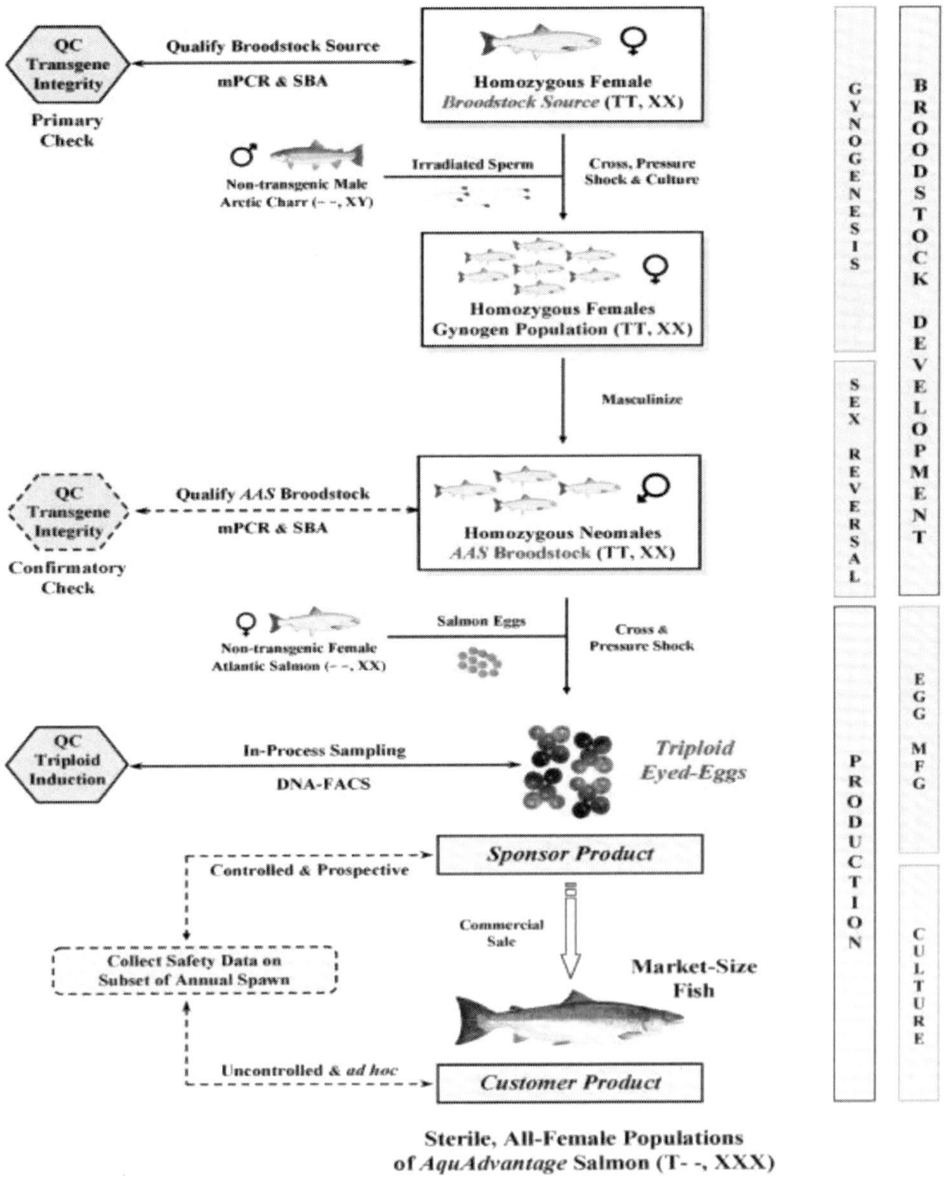

Figure 3. Graphical representation of the AquAdvantage Salmon production plan.

Conclusion: No hazards have been identified in the production plan. A possible source of uncertainty is the impact of gynogenesis on the durability of the AquAdvantage construct. The additional testing of all broodstock during the first three production cycles should be sufficient to determine the impact of gynogenesis on the genetic durability of AquAdvantage Salmon. The proposed production plan is acceptable.

b. Presence and Stability of AquAdvantage Construct

Testing for the presence of the AquAdvantage construct and verification of its location is conducted following the PCR method outlined below.

PCR Procedure

The method takes advantage of standard multiplex Polymerase Chain Reaction (PCR) using sets of primers, whose composition is based on the sequence of the rDNA construct and the regions flanking the site of integration in the AquAdvantage Salmon genome. ABT's PCR assay allows reliable differentiation between the approved GE fish and their unmodified counterparts and provides confirmation of the presence of the approved rDNA construct in the fish. Samples from both GE and unmodified fish generate two DNA amplicons corresponding to the endogenous growth hormone gene. However, samples from AquAdvantage Salmon amplify an additional DNA fragment, which is unique for the approved AquAdvantage construct. The PCR assay uses additional primer pairs to amplify DNA regions at the 5' and 3' junctions of the EO-1a insertion site and verify that the construct remains in its originally-approved location and orientation within the genome.

Reference Standards and Controls

The PCR method includes one reference standard and several controls. The specified reference standard is a commercially available 100 bp DNA ladder. Each PCR assay contains two control samples: a positive control from a known AquAdvantage Salmon and a negative control from a known non-GE salmon. The positive control produces a band consistent with the presence of the transgene in the sample. The negative control does not have the transgene, and therefore will not show this band.

Additionally, the method includes a no-template control. This assay should not show any bands unless the sample has been contaminated. Therefore, this control determines whether the entire assay has been compromised via the introduction of contaminating DNA.

Each individual reaction in the assay also contains its own internal control. The primers amplify amplicons from endogenous growth hormone in all fish irrespective of their GE status. The presence of these DNA fragments in the PCR product mixture confirms the quality of the genomic DNA and PCR conditions.

Together, these controls and standards provide evidence that the assay was conducted properly and its results are valid.

PCR Method Validation

Validation of a durability indicating method considers method accuracy, precision, specificity, linearity, range, limits of detection and quantitation, ruggedness, and whether the method is capable of detecting a durability failure. The extent to which these factors are considered is dependent on the type of method and its application.

The proposed method is a qualitative PCR assay designed to determine the presence or absence of the AquAdvantage construct in the fish and whether that construct remains at the original EO-1α site of integration. PCR, by its very nature, is reasonably accurate, precise, and specific. The relative ease of performing a PCR assay and the likelihood of its success primarily depends on the complexity of the DNA molecule being analyzed, the efficiency of the primer-DNA template hybridization, and the stringency of the PCR conditions. Provided that the primers are appropriately designed and the conditions of the reaction are specified, they will only bind to, and therefore amplify, the unique target sequence of DNA. In the event that there are mismatches between the primer and target sequences, the PCR reaction will likely not take place. The qualitative nature of the method provides a binary output provided that the conditions of the reaction are suitable – either the reaction occurred because the target sequences are present, or the reaction did not occur because the target sequence is absent or sufficiently changed so as to render the kinetics of primer binding inefficient under the method conditions. Therefore, linearity and range are not at issue for this type of method.

Limits of detection and quantitation are important measures for quantitative methodologies. Although quantitative PCR methods do exist and such considerations would be evaluated for those methods, the method at issue here is a qualitative PCR. Practical limits of detection do exist for even qualitative PCR. However, given that ABT will have ready access to sample DNA, this is not a concern for this particular method. Similarly, with respect to qualitative PCR, methodologies are extraordinarily rugged and robust under the specified conditions of the reaction. Even significant deviations from the specified conditions often have little or no effect on the result.

PCR methods are also durability indicating. For a method to be durability indicating, it should be able to detect changes in the sequence, arrangement, or location of the construct. The types of changes we are interested in are gross changes, such as large deletions, insertions, duplications, or rearrangements; more subtle changes, such as single nucleotide changes, are not *per se* durability failures. In the unlikely event that a change in the DNA sequence that is not detectible using qualitative PCR methods results in a safety or effectiveness issue, we rely on post-approval product monitoring.

The proposed method includes specific primers to be used and the conditions under which the reaction is to be conducted. The specific reaction conditions and methods for extracting the sample DNA were also provided to FDA. The primer sequences and specified reaction conditions are appropriate for determining the presence and stability of the AquAdvantage construct in AquAdvantage Salmon.

Conclusion: The qualitative PCR method is acceptable as a method for determining the presence and stability of the AquAdvantage construct at the α-locus.

c. Copy Number

The method takes advantage of the Southern blot technique to identify the number of AquAdvantage constructs and their respective integration sites in the animal genome. Southern blot analysis was performed by a contract testing laboratory. The method was conducted using GLP and current Good Manufacturing Practices (cGMP). The method is capable of distinguishing insertion of the construct at the α-locus versus other locations in the genome.

Southern Blot Procedure

Briefly, genomic DNA is extracted from blood samples using standard DNA extraction techniques. DNA quality is confirmed via agarose gel electrophoresis and ethidium bromide staining. Qualified DNA samples are digested singly with *Pst* I and *Eco*R I restriction enzymes. Digestion of each sample is confirmed by agarose gel electrophoresis. The remaining digestion mixture is resolved by electrophoresis followed by transfer onto nylon membranes. A specific DNA fragment is radioactively labeled and used to probe the nylon membrane. Autoradiograms for the *Pst* I and *Eco*R I digests are obtained by exposing film for specific periods of time.

If the construct is present at the expected α-locus, the Southern blots will show specific banding patterns in each of the two digests.

Reference Standards and Controls

The Southern Blot method includes molecular weight standards and two controls. Molecular weight standards are a commercially available mixture of *Hin*d III digested λDNA and *Hae* III digested φX174 DNA. Each autoradiogram contains two controls: a negative (non-GE salmon genomic DNA) and a positive (non-GE salmon genomic DNA combined with AquAdvantage construct). Negative control lanes are expected to reveal no bands on the film, whereas positive control lanes should exhibit the specific banding pattern consistent with appropriate insertion of the transgene construct.

Southern Blot Validation

As discussed above with respect to the PCR validation, validation of a durability indicating method considers several factors depending on the type of method and its application. The described method is intended to find all inserted copies of the construct within the salmon genome. Accuracy, precision, and specificity of a Southern blot method depend on the design of the probe and the region of genomic DNA that is being probed. The validity of the probe is confirmed during each blot through the use of the positive and negative control samples.

Similar to a PCR method, this technique is qualitative. Linearity, range, and limits of quantitation are therefore not a concern. Although not quantitative, the proposed method is stability indicating. Genomic events that would lead to a durability failure are detectable with this method, and include deletions (loss of the construct), duplications (appearance of additional copies of the construct in other genomic locations), mobilizations (movement of the construct to another site), and concatamerizations (addition of a new copy at an existing site of insertion). Each of these four events will lead to changes in the banding patterns on the exposed film.

It is important to note that although this method will be able to detect a durability failure as a result of mobilization, duplication, concatamerization and loss of the gene, it is not able to distinguish between a hemizygous (having just one AquAdvantage construct allele) and a homozygous (two copies of the allele) fish. This deficiency is not a significant regulatory concern for several reasons. First, the original hemizygous fish were bred to homozygosity using an appropriate breeding scheme. Each fish's zygosity was determined through observation of the phenotype of its out-crossed offspring; a homozygous fish would produce a population of fish who were all fast-growing whereas a hemizygous fish would produce a bimodal population with respect to growth rate. ABT maintains records of this backcrossing for each fish in its lineage. The gynogens of a homozygous fish should be homozygous; therefore it is reasonable to assume that a confirmatory Southern Blot can be interpreted to mean that the animal is homozygous.

Second, should this assumption be false, the post-market surveillance program (discussed in detail below) will identify bimodal distributions of fish at the Panamanian facility, thereby alerting ABT of a durability failure. The risk of a durability failure due to a hemizygous broodstock does not represent a safety concern. Furthermore, in the unlikely event that such a failure occurs, FDA will be notified through the post-approval reporting requirements that some of the fish did not meet the product's rapid growth claim.

Finally, although not fully validated at this time, ABT is developing a quantitative PCR method to replace the technically difficult and time-consuming Southern Blot method. This quantitative PCR method will have the same capabilities as the Southern Blot method and, in addition, will distinguish between one and two copies of the AquAdvantage construct allele.

Conclusion: The Southern blot method is acceptable as a method for determining copy number stability in the EO-1α lineage.

d. Triploidy

Triploidy is one method of producing reproductively incompetent fish for use in aquaculture and fishery management, and is widely accepted as the most effective method for this purpose.[12] An animal is triploid when it contains three copies of its genomic chromosomes, as opposed to the usual two copies. In the case of certain species of fish, triploidy results in fish that are similar in most respects to diploids when examined at the whole animal level with the exception of rendering the fish incapable of reproducing (especially in female fish, such as the AquAdvantage Salmon). The impact of triploidy on the AquAdvantage Salmon is discussed at length in the evaluation of the Phenotypic Characterization (see Section V above). ABT uses triploidy as a method for genetic containment, for both environmental risk management as well as protection of intellectual property.

ABT submitted results from a study designed to validate the process used to induce triploidy in AquAdvantage Salmon.

Induction of Triploidy

In five successive weeks, ten 1-to-1 crosses were established between 10 non-transgenic female Atlantic salmon and milt from eight AquAdvantage Salmon males hemizygous for EO-1α. Two crosses were generated in each of five successive weeks. Bulk fertilized eggs from each cross were apportioned by volume into four samples of approximately 1,500-3,000 eggs. To induce triploidy, fertilized eggs were placed in a 750 mL pressure vessel and subjected to a specific pressure for a specified period of time. Following pressure treatment, eggs were water hardened, disinfected, and incubated until sampling for ploidy count. When the treated eggs reached the eyed stage, approximately 350 eggs were arbitrarily sub-sampled to estimate triploidy rate.

Percent Triploidy Results: The % of triploid eggs from each induction was estimated and a lower 95% confidence bound was estimated using exact binomial methods rather than methods based on the assumption of normality. The pooled percentage ploidy for crosses were estimated assuming that all samples were combined. Results are provided in Table 10. The counts made from samples within a cross provide an estimate of within cross variability and the counts made for all crosses provide an evaluation of the robustness of the method.

Of the 20 samples, 14 samples had a lower 95% confidence bound greater than 99% with 19 samples having a lower 95% confidence bound greater than 98%. The estimated sample average percent of triploidy for the 5 crosses ranges from 99.7%-99.9%. For the triploidy rate estimates within a cross, the minimum range was 99.7-100% and the maximum range was 98.9-100%. The tight ranges indicate that the induction method has low variability within a given cross.

Conclusion: We have confidence that the method will provide triploid rates greater than 98% for most inductions.

Table 10. Summary analysis of triploid induction for multiple batches of multiple crosses

Cross Code	% Triploid	Lower 95% CB[1]	Average % Triploid
ABF08-AS10PSa	100	99.2	99.9
ABF08-AS10PSb	99.7	98.6	
ABF08-AS10PSc	100	99.2	
ABF08-AS10PSd	100	99.2	
ABF08-AS15PSa	99.4	98.2	99.7
ABF08-AS15PSb	100	99.2	
ABF08-AS15PSc	100	99.2	
ABF08-AS15PSd	99.4	98.2	
ABF08-AS43PSa	100	99.2	99.7
ABF08-AS43PSb	98.9	97.4	
ABF08-AS43PSc	100	99.2	
ABF08-AS43PSd	100	99.2	
ABF08-AS45PSa	100	99.2	99.9
ABF08-AS45PSb	99.4	98.2	
ABF08-AS45PSc	100	99.2	
ABF08-AS45PSd	100	99.2	
ABF08-AS54PSa	99.7	98.7	99.9
ABF08-AS54PSb	100	99.2	
ABF08-AS54PSc	100	99.2	
ABF08-AS54PSd	100	99.2	
Average for all 5 Crosses			99.8

[1] CB = Confidence Bound.

Post-Approval Sampling Plan

To evaluate the triploidy process in production batches, eggs are placed in a 3 L vessel for pressure treatment and concomitant triploidy induction. A number of such "batches" are combined in a 23 L upwelling chamber (referred to as a "lot"), which ultimately contains 100,000 to 200,000 eggs. Fertilized eggs are subject to an increasing flow of water that facilitates thorough mixing. From this chamber, pooled samples are taken for analysis and analyzed by fluorescence-activated cell sorting (FACS) analysis to determine ploidy.

The overall ability of the sampling plan to serve as an appropriate and discriminatory process control depends on the specific acceptance criteria. As an initial matter, the release testing is aimed at controlling for false positive results, namely the commercial release of eggs that, in fact, do not meet a certain minimum requirement of triploidy.

The minimum level of triploidy acceptable for AquAdvantage Salmon is driven by the Environmental Assessment (see Section VIII below), rather than other steps of the evaluation process. The Environmental Assessment considers a multiply-redundant system of both physical and biological containment. Although this ostensibly allows for flexibility in the approach toward triploidy as one biological containment strategy, a high degree of confidence with respect to triploidy provides a layer of protection against failure of other containment strategies.

The Environmental Assessment assumes that each batch will contain at least 95% triploid eggs. Therefore, the process control release specifications should be set so as to assure that with an approximate 0.95 probability at least 95% of released eggs are triploid.

The proposed sampling procedure consisted of the following four steps.

Step 1: Determine the proportion of 200 eggs that are triploid.

Step 2: If the proportion of triploid eggs ≥ p1, release the lot. Otherwise, sample 700 additional eggs.

Step 3: Determine the proportion of 700 additional eggs that are triploid and estimate proportion of all 900 eggs that are triploid.

Step 4: If the combined proportion of triploid eggs ≥ p2 release the lot. Otherwise, destroy the lot.

The release specifications used as the Step 2 and Step 4 criteria were estimated using a series of simulations (see Table 3 below). In each simulation 100,000 eggs were generated, with a specified true proportion (true p) of triploid eggs (the remainder being diploid). From each simulation 200 eggs were randomly selected (Step 1) and the proportion of triploid eggs was compared to p_1 (Step 2). If the criterion were not met, an additional 700 eggs were selected (Step 3) and the proportion of triploid eggs in the combined sample of 900 eggs was compared to p_2. Table 11 shows the proportion of the 10,000 simulations that passed the p1 and p2 criteria.

Table 11. Simulation (10,000) of the sampling plan (chamber size = 100,000 eggs)

True p	Probability of Lot Being Accepted			
	$p_1 = 0.98$	$p_2 = 0.975$	$p_2 = 0.964$	$p_2 = 0.950$
0.90	0.0000	0.0000	0.0000	0.0000
0.91	0.0000	0.0000	0.0000	0.0000
0.92	0.0004	0.0004	0.0004	0.0009
0.93	0.0010	0.0010	0.0010	0.0088
0.94	0.0070	0.0070	0.0076	0.1177
0.95	0.0261	0.0261	0.0432	0.5376
0.96	0.0925	0.0963	0.3189	0.9432
0.97	0.2796	0.3641	0.8652	0.9993
0.98	0.6257	0.9090	0.9993	1.0000
0.99	0.9500	0.9998	1.0000	1.0000

Because we are controlling for an overall 5% false positive rate that released batches contain at least a true 95% proportion of triploid eggs, we want to identify those specific criteria that meet these constraints. As a result, we identified cells in the table where the true p was 0.95 and the proportion of simulations that met the release criteria was £ 0.05. Of the simulated sampling schemes, a Step 2 criterion (p_1) of 0.98 and a Step 4 criterion (p_2) of greater than or equal to 0.964 (shaded cells above) are, therefore, appropriate.

ABT has elected a Step 2 criterion which we would expect to permit only an approximate 3% false positive rate. Specifically, ABT could have selected a lower Step 2 criterion and still maintained an acceptable false positive rate. The Step 4 criterion is expected to provide the overall false positive target rate of 5%. Therefore, although the Step 4 criterion is less stringent than the Step 2 criterion, the end result is more than adequate to address our concerns.

Conclusion: The proposed sampling plan and method for determining egg ploidy is acceptable for the qualification of production lots of AquAdvantage Salmon eggs.

e. Out-of-Specification Procedures

ABT has further committed to retesting any test samples or production lots found to be out-of-specification (OOS). All OOS results will be investigated to determine the cause of the result or, in the case of a triploidy failure, will result in destruction of a production lot. Any confirmed OOS results will result in the disqualification of that animal from the broodstock or the destruction of that production lot. Should the EO-1α lineage fall out of specification, ABT commits to procedures for the regeneration of the line, including maintenance of cryogenically preserved milt at two distinct locations.

ABT commits to annual reporting of their results of the durability testing following each production cycle.

Conclusion: The out-of-specification procedures are adequate and acceptable.

f. Post-Approval Safety Surveillance

ABT provided basic information on their plans for post-approval surveillance with respect to animal safety (i.e., mortality, morbidity, morphology) for the PEI broodstock facility. Information on animal safety surveillance to be conducted at the Panama facility was also provided in the revised Durability Plan. The proposed surveillance will involve routine assessments conducted in compliance with established inventory management practices and data collection under existing procedures (and not under a separate study protocol). Although more detailed information on the surveillance plans and consolidation of this information in the Durability Plan was not provided, the general approach described in the plan is acceptable as these procedures have been developed, refined, and implemented over a number of years.

For the PEI broodstock facility, assessments of morbidity-mortality and morphology will be conducted on the annual spawn during the early-life stages. Assessments will include evaluation of morphologic irregularities in a predetermined number of arbitrarily selected animals (fry and juveniles) prior to selection of any group of fish for grow-out or culling. The disposition of animals subject to culling, including the reason(s) for the associated reduction in inventory, will be addressed. The plan provides for collection of a balanced dataset on the incidence of irregularities for all AquAdvantage Salmon crosses established for breeding and includes assessments of all sub-populations culled. Overall, this approach addresses our primary concern that the existing data for animal safety may be biased because data were not collected on the sub-populations of fish that were culled, particularly fish culled during early-life stages, namely prior to, or soon after swim-up, when malformations might be most apparent.

Plans for collection of safety data at the grow-out facility in Panama calls for ABT to collect data on mortality-morbidity and morphologic irregularities in swim-up fry (body weight ~1-2 g) and fry-smolt (body weight ~20-100 g) prior to their transfer to grow-out tanks. The assessments will be conducted on a pre-determined, statistically-appropriate number of arbitrarily selected animals in each tank prior to selection of any group of fish for grow-out or culling, should such selection occur. The overall surveillance approach is acceptable and addresses concerns about collecting animal safety data under commercial grow-out conditions and on early-life stages of fish.

Conclusion: The proposed studies are adequate to address questions that remain from the Phenotypic Characterization step. The surveillance program design is therefore acceptable.

C. Conclusion

The information provided by ABT supports a finding that (1) both the genotype and phenotype of AquAdvantage Salmon are durable, and (2) ABT has in place an acceptable plan to ensure the future durability of the EO-1α lineage of fish.

VII. Food Safety

A. Overview

In this step of the hierarchal review process, the data and information submitted in support of a food safety assessment of triploid, all female, GE salmon containing the AquAdvantage construct was evaluated. For the purposes of this evaluation, food refers to food and feed. This step includes information and conclusions drawn from prior steps of the AquAdvantage Salmon evaluation, as well as data and information evaluated for the identity, composition, level(s) of expression product from the rDNA construct, and other potential downstream hazards that may be influenced by the expression product, and allergenicity. Food Safety Appendix A includes the evaluation of the analytical method used as the regulatory method. The evaluation meets FDA's statutory requirements for demonstrating food safety as described by Guidance for Industry 187: *Regulation of Genetically Engineered Animals with Heritable Recombinant DNA Constructs*, this evaluation is consistent with the Codex Alimentarius Commission's *Guideline for the Conduct of Food Safety Assessment of Foods Derived from Recombinant-DNA Animals* (CAC, 2008).

The subject of this new animal drug application was a triploid, monosex (all female) Atlantic salmon containing the AquAdvantage construct at the α-locus; these animals are referred to as "AquAdvantage Salmon." Some of the data sets evaluated also include data and information on diploid male and female GE salmon as well as triploid female GE salmon of the same lineage; these are referred to as ABT salmon. The results of those analyses are also included.

The primary risk question associated with food consumption is whether there are any risks of direct or indirect effects associated with the consumption of edible products derived from this GE salmon. The conclusions of this assessment are provided in the context of food safety; that is, because no food is completely safe, the most appropriate way in which to consider the risk question is to determine whether there is any difference between food from ABT salmon and other Atlantic salmon, or whether food from ABT salmon is as safe as food from other Atlantic salmon.

To this end, we conducted a weight-of-evidence evaluation of the data and information provided in support of a food safety assessment. After evaluating all of the data and information, including peer-reviewed publications and publicly available data bases as appropriate, we concluded the following:

- ABT salmon meet the standard of identity for Atlantic salmon established by FDA's Reference Fish Encyclopedia.
- Food from AquAdvantage Salmon (triploid, monosex (all female) ABT salmon) is the same as food from other Atlantic salmon.
    - No biologically relevant differences were detected in the levels of the gene product (the Chinook salmon growth hormone), or any endogenous metabolite or substance found in physiological pathways that could be impacted by that hormone (the somatotropic axis);
    - No biologically relevant differences were noted in either the gross composition (proximate analysis), or in any edible tissue component (e.g., amino acids, minerals, vitamins, fatty acids).
        - ABT salmon contained the expected amounts of nutritionally important omega-3 and omega-6 fatty acids at the appropriate ratio for a fish source.
    - No biologically relevant differences were found in the allergenicity of edible products of AquAdvantage Salmon.
- We have found no biologically relevant difference between food from ABT salmon and conventional Atlantic salmon based on the criteria evaluated.
- No direct or indirect food consumption hazards were identified in AquAdvantage
- Salmon (triploid, monosex (all female) ABT salmon).
- Food from AquAdvantage Salmon (triploid, monosex (all female) ABT salmon) is as safe to eat as food from other Atlantic salmon.
- We have a high degree of certainty in our conclusions regarding AquAdvantage
- Salmon (the triploid, monosex (all female) ABT salmon).
- The analytical method of identity for AquAdvantage Salmon is acceptable. We determined that there was no need to set a tolerance for the residues resulting from the insertion of the construct to produce AquAdvantage Salmon.

We therefore conclude the food from AquAdvantage Salmon (the ***triploid*** ABT salmon) that is the subject of this application is as safe as food from conventional Atlantic salmon, and that there is a reasonably certainty of no harm from the consumption of food from this animal. No animal feed consumption concerns were identified.

The basic risk question for the consumption of food derived from GE animals is "What is the risk of *direct* or *indirect* effects associated with consumption of edible products derived from the GE animal?"

Because no food is completely safe, the most appropriate way to approach the risk question above is to ask whether there are any differences between food from the AquAdvantage Salmon and other Atlantic salmon that pose a food consumption risk. In other words, will food from AquAdvantage Salmon be as safe as food from other Atlantic salmon?

Direct effects, for the purposes of this food safety evaluation, are defined as those that arise from consumption of edible products from the GE animal, including exposure to the AquAdvantage construct or its gene product (i.e., the Chinook salmon growth hormone). Because nucleic acids, including DNA, are presumed to be Generally Recognized As Safe (FDA, 1992), there is no direct food consumption risk associated with exposure to the AquAdvantage construct itself. Evaluation of direct food consumption effects is therefore limited to those associated with exposure to the Chinook salmon growth hormone in food from the AquAdvantage Salmon.

Indirect effects, for the purposes of this evaluation, are those effects that can be attributed to the rDNA construct or its gene product perturbing the physiology of the animal. These could alter the composition of food and may pose an increased risk compared to consumption of food from the appropriate comparator.

The distinction between direct and indirect effects is not always clear-cut, and it may be that the evaluation of certain lines of inquiry can fit into either or both categories. For the purposes of this evaluation we have made the distinctions found in Table 12.

Table 12. Characterization of Potential Direct and Indirect Effects

Direct Effects	Indirect Effects
• Alterations from Gene Expression Product (Chinook salmon growth hormone) including • Alterations in Levels of Hormones Associated with the Somatotropic Axis, including IGF1 • Allergenicity of the Gene Expression Product	• Alterations in the Composition of Edible Tissues • Alterations in the Endogenous Allergenicity of Edible Tissues

B. Is AquAdvantage Salmon an Atlantic Salmon?

In order to conduct analyses to determine whether changes in composition at either the gross or fine level occurred as the result of the introduction of the AquAdvantage construct, or if AquAdvantage Salmon was more allergenic than other Atlantic salmon, we first needed to determine whether AquAdvantage Salmon was indeed Atlantic salmon.

The essential nature of the salmon has not changed as a result of the introduction of the AquAdvantage construct and, as a result, the AquAdvantage Salmon is still Atlantic salmon (see, for example, Memorandum from Linda S. Kahl to Acting Director, Office of Premarket Approval, Center for Food Safety and Applied Nutrition, Food and Drug Administration. May 17, 1994).

An empirical confirmation that AquAdvantage Salmon is, in fact, an Atlantic salmon can be accomplished by referring to the FDA Regulatory Fish Encyclopedia (RFE). The RFE is a searchable compendium of some 1,700 species of fin- and shell-fish developed by FDA scientists at the Seafood Products Research Center (SPRC, Seattle District), and the Center for Food Safety and Applied Nutrition (CFSAN) to help federal, state, and local officials and purchasers of seafood identify species substitution and economic deception in the marketplace (Available at http://www.fda.gov/Food/FoodSafety/Product-Specific Information/Seafood/RegulatoryFishEncyclopediaRFE/default.htm).

Isoelectric Focusing (IEF) gel banding patterns have been developed for 57 specimens from 39 species within the RFE, to provide a chemical taxonomy based on species characteristic biochemical patterns that may be used in species identification. The following FDA study evaluated ABT salmon tissue using the RFE standardized approach.

Comparison *of Growth-Hormone Transgenic Fish Atlantic Salmon Salmo salar Edible Tissue with the FDA/CFSAN RFE Standard for Non Transgenic Fish. FDA/CVM Office of Research. Report dated 3 December 2004.*

The goal of this study was to determine whether there were differences in the IEF and 2-dimensional gel (2D gel) electrophoresis fingerprints between non-GE Atlantic salmon and ABT salmon. PCR was used to verify the presence or absence of the AquAdvantage construct, based on a probe sequence obtained from ABT.

ABT salmon and non-GE Atlantic salmon samples were obtained as blinded samples from ABT and stored at -80°C. ABT also provided identified samples to use as controls. Reference samples of Atlantic salmon were purchased from a local market and served as internal controls. These samples were also stored at -80°C.

Protein was extracted from the samples and the extracts were used for IEF and 2D gel analysis. Sample identification and the presence of the AquAdvantage construct were confirmed by ABT. The IEF and 2D gel results showed no appreciable differences in banding patterns (see Food Safety Appendix B). The finding of identical IEF banding patterns confirmed that the ABT salmon, and thus, AquAdvantage Salmon met the standard of identity for Atlantic salmon under the criteria developed for the RFE.

Some differences were noted in the intensity of some of the bands on the 2D gels among all tissues assayed, including within controls. This likely indicates differences in gene expression among the salmon samples. This is expected, as slight gene expression changes within individual animals is a consequence of natural genetic and epigenetic variations. Based on this rationale, these differences in banding intensity are not relevant to the confirmation of the identity of the ABT/AquAdvantage Salmon as Atlantic salmon.

Conclusion: AquAdvantage Salmon meet FDA's standard for identity for Atlantic salmon under the criteria established for the RFE.

C. Hazard Characterization for Food Safety Determination

1. Characterization of Direct Food Consumption Hazards

Based on the hazard identification steps, the only direct hazards identified were those related to the expression product of the introduced construct (i.e., Chinook salmon growth hormone) and endogenous substances in the salmon that could be altered as the result of changes in growth hormone expression (i.e., IGF1). In order to characterize this potential hazard, ABT contracted studies to measure the levels of the Chinook growth hormone and select other hormones in GE salmon and non-GE comparator salmon.

a. Analysis of Gene Expression Product

The following section begins with a discussion of information on the gene expression product available from peer-reviewed journals, and continues with studies performed by ABT to address this particular issue. The peer-reviewed study is addressed first because it provides a framing for consideration of potential hazards that may be found in AquAdvantage Salmon.

i. Peer-Reviewed Publication: Du, S.J., A. Gong, G.L. Fletcher, M.A. Schears, M.J. King, D.R. Idler, and C.L. He (1992) Growth Enhancement in Transgenic Atlantic Salmon By The Use Of An "All Fish" Chimeric Growth Hormone Gene Construct. Nature Biotechnol. 1:176.

The developers of the lineage of growth hormone (GH) transgenic Atlantic salmon that became the AquAdvantage Salmon published a report in 1992 in which selected plasma hormone levels were measured. The GH transgenic Atlantic salmon used in this study were derived from the same parental animals from which the EO-1α lineage was eventually derived. This study reported on 500 GE and non-GE Atlantic salmon, resulting from Atlantic salmon eggs injected with the ocean pout antifreeze protein (AFP) promoter-salmon growth hormone (GH) construct in November, 1989. The 500 Atlantic salmon were selected for PCR analysis with the 200 heaviest (largest) GH transgenic Atlantic salmon selected. All salmon weighing more than eight grams (n=14) had blood sampled and were tagged; 36 additional fish (weighing more than five grams) were also bled and tagged. Of these total 50 fish, six were shown to contain the construct in both their red blood cells and in their scales. (One additional fish appeared to be a mosaic, containing the construct in its scales, but not in its red blood cells, and was eliminated from further analysis). Body weight and plasma levels of growth hormone and triiodothyronine (the thyroid hormone often referred to as T3) were determined. Control salmon derived from non-injected eggs of approximately the same age were also weighed and plasma concentrations of the same two hormones were analyzed.

Body weights and plasma concentrations of growth hormone and T3 measured on January 12, 1991 are abstracted from Table 1 of the publication and provided below in Table 13. Mean plasma growth hormone concentrations did not differ statistically between the GH transgenic Atlantic salmon and either comparator (non-GE siblings or control samples). Mean plasma T3 concentration in the GH transgenic Atlantic salmon was statistically different from their non-GE siblings, but not statistically different from the farm control salmon. The authors noted that plasma GH levels ranged from 9.5 to 91.4 ng/ml in the GH transgenic Atlantic salmon, with the largest salmon having the lowest concentrations. Similarly, the three largest GH transgenic Atlantic salmon had much lower plasma T3 concentrations (0.26 ± 0.18 ng/ml compared to the two smaller GE Atlantic salmon (2.34 ± 0.14 ng/ml)).

The authors noted that GH enhances conversion of thyroxine (T4) to triiodothyronine (T3) in eels, and speculated that a similar action may occur in salmon. They speculated that there may be a negative feedback loop between GH and T3 which diminishes T4 production, and proposed that decreased T3 levels may serve as an indicator of expression of the GH from the inserted AquAdvantage construct. Regardless of the mechanism, the authors concluded that there was no correlation between blood plasma GH levels and growth rates or presence of the transgene (rDNA construct). The authors further speculated on the mechanism by which the expression of the construct mediates increased growth. They proposed that the arctic pout promoter provides a tissue specific expression in the liver, facilitating more efficient interaction between GH and the GH receptors in liver cells. This, in turn, may permit very low GH expression (with little or no increase in plasma concentration) to stimulate expression of IGF1, resulting in increased growth. Some support for this mode of action was provided in the 2008 study by Hobbs and Fletcher. In this study, the authors were able to show low level mRNA expression of the inserted construct in the EO-1α lineage in a wide selection of tissues.

ii. ABT's Study: Determination of IGF1, GH, T3, T4, 11-Keto Testosterone, Testosterone, and Estradiol in Salmon Tissue. CTBR Bio-Research Inc. Canada. Project Number 42361. Study Report AAS-HFS-001. Report dated 26 July 2004.

The purpose of this GLP compliant study was to determine the concentration of insulin like growth factor 1 (IGF1), growth hormone (GH), estradiol, testosterone, T3, T4, and 11-ketotestosterone in salmon muscle and skin.

Table 13. Body Weight and Plasma Concentrations of Growth Hormone and T3

	GH (ng/ml)	T3 (ng/ml)	Fish Weight (g)
GH Transgenic Atlantic salmon	39.9 ± 14.8 (5)	1.1 ± 0.5 (5)	*47.3 ± 9.5 (6)*
Non-GE siblings & P-value	28.2 ± 8.8 (7) NS	2.8 ± 0.5 (5) <0.05	*9.48 ± 0.6 (43) <0.01*
Controls & P-value	20.5 ± 7.97 (5) NS	1.9 ± 0.1 (3) NS	*10.4 ± 0.6 (10) <0.01*

Growth hormone (GH) and triiodothyronine (T3) measured from the five largest GH transgenic Atlantic salmon, the largest non-GE siblings, and five blood samples pooled from two control salmon derived from non-injected eggs. Values presented as mean ± standard error. Statistical comparisons were made between the GH transgenic Atlantic salmon and their non-GE siblings or the controls (method of analysis not reported).

Tissue samples were taken from a total of 73 diploid and triploid market-size salmon (10 farmed control, 33 sponsor control, and 30 ABT salmon). Validation information for each of the assay methodologies, and calibration data for the performance of the actual assays were provided. Calibration curve parameters, calculated concentrations for the standards, and calculated concentrations for the quality control samples were provided.

Table 14 provides the summary of units, limits of quantitation and limits of detection for the assays. The mean concentration of IGF1, growth hormone, estradiol, testosterone, T3, T4, and 11-keto testosterone for the farm control, sponsor control, and "treated" (i.e., GE) fish (each respective group pooled for gender and ploidy) are summarized in Table 15.

Growth hormone was below the limit of quantitation in all samples, whether in the treated (GE), sponsor control, or farm control groups. Treated (GE) salmon did not have statistically different concentrations of estradiol, testosterone, 17- ketotestosterone, T3, or T4 when compared to sponsor control fish. Initial evaluation of the results suggested that there may have been an increase in the level of IGF1 in the GE fish compared to sponsor control fish. A further evaluation of the data showed that the most apparent potential difference were between the mature diploid sponsor control and the mature diploid GE salmon. The individual values are reproduced in Table 16.

Table 14. Assay Parameters for Selected Hormones in Salmon Tissue

Analyte	Tracer	Assay parameters*				Amount per gram of tissue (ng/g)*	
		Concentration in assay specific units		Concentration (ng/mL)			
		$LLOQ_{assay}$	$ULOQ_{assay}$	$LLOQ_{assay}$	$ULOQ_{assay}$	$LLOQ_{tissue}$	LOD_{tissue}
IGF1	125I-IGF1†	1,500.0	50,000.0	1.500	50.00	3.27	2.18
GH	125I-GH†	2,500.0	20,000.0	2.500	20.00	10.40	6.24
Estradiol	125I-estradiol†	17.5	1,800.0	0.0175	1.80	0.018	a
Testosterone	125I-testosterone‡	36.4	1,018.3	0.459	10.18	0.46	a
T3	125I-T3‡	36.4	584.0	0.364	5.84	0.36	a
T4	125I-T4#	1.5	15.0	15.000	150.00	15.00	a
11-keto testosterone	11-keto testosterone-acetylcholin- esterase†	18.9	850.2	0.019	0.85	0.019	a

* LLOQ=l of quantitation, ULOQ=upper limit of quantitation, LOD=limit of detection
† Assay units are pg/mL.
‡ Assay units are ng/dL.
\# Assay units are μg/dL.
a = not determined.

Table 15. Summary of Contractor's Hormone Analysis

Variable	Group	*N*	mean	std	min	max
Estradiol	GE	20	0.36	0.375	0.02	1.32
	SControl	22	0.38	0.439	0.02	1.85
	FControl	5	0.04	0.019	0.02	0.06
Growth hormone	GE	0	<LOQ	n/a	n/a	n/a
	SControl	0	<LOQ	n/a	n/a	n/a
	FControl	0	<LOQ	n/a	n/a	n/a
IGF1	GE	6	10.26	4.971	3.97	18.43
	SControl	11	7.34	2.818	3.56	12.24
	FControl	0	<LOQ	n/a	n/a	n/a
11-keto testosterone	GE	29	86.21	92.490	20.76	389.52
	SControl	33	71.42	87.302	21.00	380.53
	FControl	10	55.27	30.357	23.17	101.97
T3	GE	26	0.85	0.312	0.44	1.59
	SControl	28	0.84	0.270	0.41	1.57
	FControl	10	1.31	0.505	0.73	2.01
T4	GE	2	19.65	0.426	19.35	19.95
	SControl	2	19.96	3.746	17.32	22.61
	FControl	2	18.52	1.320	17.58	19.45
Testosterone	GE	25	1.06	0.476	0.46	2.21
	SControl	30	1.17	0.692	0.55	3.35
	FControl	10	1.01	0.646	0.52	2.68

N = number of fish sampled with values above LOQ
SControl = Sponsor control (non-GE fish)
FC = Farmed fish control (non-GE fish)

Table 16. IGF1 Levels in Mature Diploid Salmon
Individual results (ng/g) (LOQ = 3.27 ng/g)

Sponsor control	6.191	6.980	7.642	8.784	9.485	10.928	12.235
GE salmon	<LOQ	3.971	6.350	10.527	10.718	11.578	18.428
	Summary Statistics						
	N*	**Mean**	**Std. Dev.**	**Min > LOQ**	**Max**		
Sponsor control	7	8.892	2.167	6.191	12.235		
GE salmon	7	9.263	5.251	3.971	18.428		

*concentrations below the LOQ were included as the LOQ value

Although there did not appear to be a statistically significant difference between the mean IGF1 level for the GE and non-GE salmon, the range of values for the GE salmon exceeded that of the non-GE salmon by more than 10%. One possible explanation of the difference in concentrations could be differences in body weights of the sampled fish with constant hormone levels. This did not appear to be the case, as body weights were shown to be similar across groups as seen in Table 17.

Table 17. Range of Body Weights

Group	**Range of Body Weights (g)**
Farm control diploid Atlantic salmon	3,972 - 5,786
Farm control triploid Atlantic salmon	3,938 - 6,604
Sponsor control diploid Atlantic salmon	2,748 - 6,896
Sponsor control triploid Atlantic salmon	2,133 - 4,286
GE diploid Atlantic salmon	2,867 - 5,813
GE triploid Atlantic salmon	2,061 - 5,865

As part of the heuristic method applied to assessing data and information, our initial decision to begin assessing the biological relevance of any measurement began with determining whether that measurement exceeded the comparator range by 10% or more[13]. Because the IGF1 levels of the mature diploid GE salmon results exceeded the IGF1 levels of the mature diploid sponsor control salmon results by more than 10%, in order to determine whether there was of biological relevance to these apparent differences, we conducted a margin of exposure assessment (MOE).

iii. Margin of Exposure for IGF1

IGF1 is an endogenous hormone that is closely linked with growth hormone expression and circulating levels (Frost and Lang, 2003). It has been considered as a potential hazard for human consumption following increased growth hormone levels in food producing animals (Juskevich and Guyer, 1990; USFDA, 1993). Although growth hormone levels were not shown to be different in the ABT salmon compared to non-GE fish, in order to ensure that the other potentially hazardous constituents along the somatotropic axis (i.e., IGF1) were not sufficiently elevated to constitute a food consumption hazard, we performed a margin of exposure assessment (MOE).

MOE assessments are often performed to determine whether exposures to a particular substance or component of the food(s) under consideration fall within the range of daily exposures, are different from those in the comparator group, and if so, whether the difference is expected to result in an adverse outcome.

Margin of exposure assessments are best performed considering both maximum likelihood and plausible upper-bound estimates of exposure. Maximum likelihood estimates consider central tendencies of intake estimates (i.e., medians or means), while plausible upper-bound limits often take the form of 95th percentile intake estimates. Both are useful in coming to conclusions regarding population exposures and characterizing the potential for substances in food to pose hazards.

In general, for purposes of this assessment, conservative (health protective) assumptions and defaults were used when data were lacking, or where inferences regarding direct or proportional intake needed to be made.

For example, because there are no reliable data on the intake of Atlantic salmon, and no GH transgenic Atlantic salmon have been marketed, one of the key assumptions we made in the initial MOE evaluation was that all of the fish consumed were Atlantic salmon, and that all of those salmon were ABT salmon. We also assumed that all of the salmon consumed contained IGF1 at the maximum concentration identified in the one outlier mature diploid animal presented in Table 16. Subsequent analyses also considered less than the upper-bound estimates by using less conservative assumptions.

The results of these analyses are found in Table 18; narrative descriptions of the information, data, and assumptions used follow immediately.

Daily human consumption of non-tuna finfish has been estimated to be 300 g per day for the 95th percentile eaters of finfish (2002).We have made the conservative (health protective) assumption that all of the finfish consumed are salmon, and adjusted that consumption value for the fraction of salmon consumed estimated to be Atlantic salmon, (approximately 2/3) 200 g per day (Knapp et al., 2007).

The upper bound for IGF1 consumption may then be estimated assuming that all salmon contain the maximum tissue levels detected in the mature diploid sponsor control Atlantic salmon and mature diploid ABT salmon.

The incremental increase calculated from the difference in residue concentrations between the mature diploid ABT salmon and the mature diploid sponsor control Atlantic salmon may also be determined. The margin of exposure between IGF1 in non-GE Atlantic salmon and diploid ABT salmon are presented in Table 19.

Comparisons to other sources of IGF1 from other finfish or food producing animals are also appropriate to contextualize this analysis. This information is summarized in Table 18.

In estimating the "natural background" levels of IGF1, we chose teenaged boys as the most "sensitive" population based on their biological sensitivity to the effects of IGF1 due to their rapid growth and development, and their tendency to consume adult portions of food despite a lower body weight (Ungemach, 1998)[14].

The results of this analysis are found in Table 19.

Table 18. IGF1 levels in Various Foods

Species	Source (tissue)	units	Range	Mean
Chinook salmon[1]	Plasma	ng/ml	5-35	-
Coho salmon[2]	Plasma	ng/ml	7-13	-
Coho salmon[3]	Plasma	ng/ml	10-15	-
Gilthead Bream[4]	Plasma	μg/L	36-100[5]	-
Bovine[6]	Raw milk	ng/ml	Intentionally Blank	5.6 ± 0.56
Bovine[6]	Pasteurized milk	ng/ml	Intentionally Blank	8.2 ± 0.35
Bovine[6]	Raw bulk milk	ng/ml	1.27-8.10	4.32 ± 1.09
Homo sapiens[6]	Milk	ng/ml	1 d post partum 17.6 2 d 12.8 3 d 6.8 6-8 wk 13-40	19
Chum salmon[7]	Plasma	ng/ml	Depends on maturity/sex/ month: varies between 16.5 and 100	-
Rainbow trout (O.kiss)[8]	Plasma	ng/ml	Function of temperature/ time Lowest value 11.2 Highest 33.6	-
Japanese beef cattle[9]	Plasma	ng/ml	Intentionally Blank	Preweaning 11.7± 3.6 Postweaning 50.5 ± 2.1
Homo sapiens[10,11]	Plasma	nmol/L	Meat-eaters 29.3-32.7 Vegetarians 29.5-32.9 Vegans 25.5-28.6	-
Polish Holstein[12]	Plasma	ng/ml	698-1024	Intentionally Blank

[1] Beckman B.R., K.D. Shearer, K.A. Cooper, and W.W. Dickhoff (2001) Relationship of insulin-like growth factor-I and insulin to size and adiposity of under-yearling Chinook salmon. *Comp. Biochem. Physiol.* Part A 129:585.

[2] Shimizu M., P. Swanson, and W.W. Dickhoff (1999) Free and Protein-bound insulin-like Growth Factor-I and IGF-binding proteins in plasma of Coho salmon. *Gen. Comp. Endocrinol.* 115:398.

[3] Pierce A.L., B.R. Beckman, K.D. Shearer, D.A. Larsen, and W.W. Dickhoff (2001) Effects of ration on somatotropic hormones and growth in Coho salmon. *Comp. Biochem Phys.* 128: 255.

[4] Perez-Sanchez J., H. Marti-Palanca, and S.J. Kaushik (1995) Ration size and protein intake affect circulating Growth Hormone concentration, Hepatic Growth Hormone binding and plasma Insulin-like Growth Factor-I immunoreactivity in a marine teleost, the Gilthead sea bream. *J. Nutr.* 125:546.

[5] Fish was fed several experimental diets. No word on how these diets are different from a standard commercial diet.

[6] Juskevich, J.C. and C.G. Guyer. (1990) Bovine Growth Hormone: Human Food Safety Evaluation. *Science* 249:875.

[7] Onuma T.A., K. Makino, H. Katsumata, B.R. Beckman, M. Ban, H. Ando,M.A. Fukuwaka,T. Azumaya, P. Swanson, and A. Urano (2010) Changes in the plasma levels of insulin-like growth factor-I from the onset of spawning migration upstream migration in chum salmon. *Gen. Comp. Endocrinol.* 165:237.

[8] Gabillard J.C., C. Weil, P.Y. Rescan, I. Navarro, J.Gutiérrez, and P.Y. Le Bail (2003) Effects of environmental temperature on IGF1, IGF2, and IGF type I receptor expression in rainbow trout. *Gen. Comp. Endocrinol.*133:233.

[9] Suda Y, K. Nagaoka, K. Nakagawa, T. Chiba, F. Yusa, H. Shinohara, A. Nihei, and T Yamagishi (2003) Change of plasma insulin-like growth factor-1 concentration with early growth in Japanese beef cattle. *Animal Sci J.* 74:205.

[10] Allen, N.E., P.N. Appleby, G.K. Davey, R. Kaaks, S. Rinaldi, and T.J. Key (2002) The associations of diet with serum insulin-like growth factor I and its main binding proteins in 292 women meat-eaters, vegetarians and vegans. *Cancer Epidemiol. Biomarkers Prev.* 11:1441.

[11] Crowe F.L., T.J. Key, *et al.* (2009) The association between diet and serum concentrations of IGF-I, IGFBP-1, and IGFBP-3 in the European Prospective investigation into cancer and nutrition. *Cancer Epidimiol Biomarkers Prev.* 18:1333.

[12] Maj A. and M. Snochowski, E. Siadkowska, B. Rowinska, P. Lisowski, D. Robakowska-Hyzorek, J. Oprzadek, R. Grochowska, K. Kochman, and L. Zwierzchowski (2008) Polymorphism in genes of growth hormone receptor (GHR) and insulin-like growth factor-1 (IGF1) and its association with both the IGF1 expression in liver and its level in blood in Polish Holstein-Friesian cattle. *Neuro. Endocrinol. Lett.* 29:981.

Table 19. Margin of Exposure Estimates of IGF1 in ABT Salmon

Description	Calculation	Value/Outcome
Daily non-tuna finfish consumption for the 95th percentile eater[1]	-	300 g/day
Assuming all consumed non- tuna finfish are salmon, consumption corrected for fraction of Atlantic salmon	2/3 * 300g/day = 200 g/d	200 g/d
Upper bound of IGF1 residue found at the maximum level in sponsor control mature diploid Atlantic salmon	-	12.235 ng/g of muscle
Upper bound IGF1 residue in mature diploid ABT salmon	-	18.428 ng/g of muscle
Daily consumption of IGF1 from Atlantic Salmon	200 g/d * 12.235 ng/g =2,447 ng/d ~ 2.4 μg/d	2.4 μg/day
Daily consumption of IGF1 from ABT salmon	200 g/d *18.428 ng/g =3,686 ng/d ~ 3.7 μg/d	3.7 μg/day
Incremental increase in daily consumption of IGF1I	3.7 μg/d – 2.4 μg/d =1.2 μg/d	1.2 μg/day
Calculated total serum burden of IGF1 in 50 kg teenager[3]	1,220,000 ng = 1,220 μg	1,220 μg/person
Margin of Exposure (MOE)	Total serum burden/ Dietary contribution	-
MOE for mature diploid (non GE) Atlantic salmon	1,220 μg/2.4 μg/day = 508	508-fold (fractionally 2.0 x 10^{-3})
MOE for mature diploid ABT salmon	1,220 μg/3.7 μg/day =330	330-fold (fractionally 3.0 x 10^{-3})

[1] US Department of Commerce (USDOC), Technology Administration, NTIS, Springfield, VA 22161 (2002) Foods Commonly Eaten in the United States. Quantities Consumed per Eating Occasion and In A Day, 1994-1996. PB2005110468.

[2] Knapp, G., C.A. Roheim, and J.L. Anderson (2007) The Great Salmon Run. Competition between Wild and Farmed Salmon. TRAFFIC North America. World Wildlife Fund. Washington, DC.

[3] Ungemach, F.R. (1998) Recombinant Bovine Somatotropins (addendum). Toxicological Evaluation of Certain Veterinary Drug Residues in Food. IPCS WHO. WHO Food Additive Series 41.

Results

We estimated by calculation the upper bound dietary exposure to IGF1 via the consumption of Atlantic salmon and ABT salmon, to be 2.4 and 3.7 μg per day, respectively.

The MOE for dietary consumption of **Atlantic salmon (non-GE)** was 1,220 μg/2.4 μg per day, which yielded a **508-fold margin of exposure**, equivalent to approximately two one-thousandths (0.002) of the total serum burden. The MOE for dietary consumption of IGF1, assuming that IGF1 was present at the **maximum concentration recorded from the mature diploid ABT salmon** cohort is 1,220 μg /3.7 μg/d, yielding a **330-fold margin of exposure,** which corresponded to approximately 0.003 of the total serum burden.

Finally, calculation of the incremental increase in IGF1 exposure from the maximum estimated GE salmon intake relative to IGF1 exposure from the study comparator yielded only 1.2 μg per day or 0.001 of the total serum burden.

Conclusions: Only growth hormone and other hormones associated with the somatotropic axis (IGF1, estradiol, testosterone, 17-ketotestosterone, T3, and T4) were identified as potential hazards for the consumption of ABT salmon in food.

Du *et al.* demonstrate that plasma growth hormone concentrations did not differ statistically between the GH transgenic Atlantic salmon (genetically engineered to contain the same construct as the ABT salmon) and either age-matched non-GE siblings or pooled control samples from age-matched siblings. Mean plasma T3 concentrations in the GH transgenic Atlantic salmon were statistically different from and lower in the GH transgenic Atlantic salmon compared to non-GE siblings but not when compared to the pooled controls. The highest plasma GH levels correlated with the largest GH transgenic Atlantic salmon while an inverse correlation was shown for mean plasma T3 concentrations.

Mean levels of estradiol, testosterone, 17-ketotestosterone, T3, and T4 were not different in the ABT salmon compared with comparator non-GE salmon (Table 4). The apparent difference in IGF1 in mature diploid ABT salmon compared to sponsor control non-GE salmon was relatively small. No differences were observed in levels of growth hormone in edible tissues at the level of quantitation for the analytical method.

Further, even if there were increases in the amounts of these normally occurring substances, Chinook or Atlantic salmon growth hormone, they would not likely effect any biologically meaningful interactions with human growth hormone receptors due to interspecies ***differences*** in the ability of these substances to bind to homologous receptors in mammals or to cause physiological changes via such binding. This lack of biological interaction is likely to also be true for mammalian and avian food producing species that could possibly eat animal feed made up of ABT salmon (see Food Safety Appendix C).

We conclude that even if the expression of IGF1 were present at the highest levels measured, and even if expected high consumers of salmon ate nothing but ABT salmon containing this likely upper ***bound*** level of IGF1, the margin of exposure to this endogenous component of food would be well within levels of exposure from other dietary sources of salmon, and poses no additional risk.

b. Potential Allergenicity of Gene Expression Product

There are two major questions with respect to the allergenicity of food from an rDNA organism: (1) the potential allergenicity of the newly expressed protein(s) present in the food, and (2) the endogenous allergenicity of the food from the recipient organism. The potential allergenicity of the gene expression product, Chinook salmon growth hormone, was considered below, as it qualified as a direct food consumption hazard. The allergenicity of food from the recipient organism was addressed under indirect food consumption hazards.

The Codex rDNA Animal Guideline describes a conservative (health protective) approach to determining whether a newly expressed protein present in a food from an rDNA organism is likely to pose an allergenic risk. This assessment strategy includes three main components including the following:

- allergenicity of the gene source;
- structural similarity to known allergens; and
- resistance to proteolytic degradation (2008).

i. Consideration of Allergenicity of Gene Source

In general, the initial step in assessing the allergenicity of a newly expressed protein in a food from a GE organism involves information regarding the history of allergic reaction of humans to the source (i.e., organism) from which the transferred gene is isolated. Transfer of a gene from an allergenic source has the potential to create a new risk for those individuals allergic to the gene source because those individuals may experience allergic reactions to foods from the GE organism which contain the transferred gene in addition to allergic reactions to the allergenic source and products derived from it (Nordlee et al., 1996).

In the case of ABT salmon, the introduced growth hormone gene was isolated from Chinook salmon (*Oncorhynchus tshawytscha*). Chinook salmon are finfish, and finfish are one of the eight major allergenic foods in the U.S. (FALCPA, 2004; Hefle et al., 1996). Although salmon growth hormones have not to-date been identified as allergenic proteins, each allergenic food contains multiple allergenic proteins, many of which have not been identified or fully characterized (Gendel, 1998). Therefore, we made the conservative (health protective) assumption that the transferred Chinook growth hormone was a putative salmon allergen. It is important to note, however, that individuals allergic to Chinook salmon also would likely be allergic to Atlantic salmon (*Salmo salar*). Because salmon present a hazard to salmon-allergic individuals, salmon-allergic individuals will likely avoid consumption of all salmon, including AquAdvantage Salmon.

ii. Analysis of Structural Similarity of Gene Product to Known Allergens

In general, approaches to assessing the allergenicity of a newly expressed protein in food from a GE organism recommend comparing the structure of the gene product to that of known allergens in order to evaluate potential IgE cross-reactivity. The Codex rDNA Animal Guideline provides guidance on how to conduct protein database searches to detect any significant amino acid sequence homologies, defined as greater than 35% identity in a segment of 80 or more amino acids, or other scientifically justified criteria (CAC, 2008). The Codex rDNA Animal Guideline also discusses searching stepwise contiguous identical amino acid segments as these may represent linear IgE-binding epitopes.

To evaluate the potential cross-reactivity of the Chinook salmon growth hormone with known allergen protein sequences, we conducted searches of the AllergenOnline database version 10 (released January 2010) (www.allergenoline.org) and the Structural Database of Allergenic Proteins (updated April 16, 2010) (www.fermi.utmb.edu/SDAP/) using deduced peptide sequences from GenBank (protein ID AAT02409.1 and AAW22586.1). These searches revealed no amino acid sequence identities of greater than 35% in segments of 80 amino acids with any entries in either database. In addition, there were no matches of eight or more contiguous amino acids with any entries in either database.

iii. Resistance to Proteolytic Degradation

A number of food allergens have been shown to be resistant to degradation by pepsin (Astwood et al., 1996). Because of this correlation between allergenic potential and resistance to pepsin digestion, newly expressed proteins in foods from rDNA organisms are typically assessed for resistance to pepsin.

For ABT salmon, the newly expressed protein is the native Chinook salmon growth hormone. There is no scientific rationale to suggest an altered resistance to pepsin when the protein is expressed in Atlantic salmon rather than in Chinook salmon. For this reason, we found the pepsin resistance assay to be unnecessary.

Conclusion: The expression of Chinook salmon growth hormone in ABT salmon does not present a new risk of allergic reaction to salmon allergic individuals and is unlikely to cause allergic cross-reactions.

c. Summary of and Conclusions from the Identification and Characterization of Direct Food Consumption Hazards

Only growth hormone and other hormones associated with the somatotropic axis (IGF1, estradiol, testosterone, 17-ketotestosterone, T3, and T4) were identified as potential hazards for the consumption of ABT salmon in food.

Du et al. demonstrated that mean plasma growth hormone concentrations did not differ statistically between the GH transgenic Atlantic salmon (genetically engineered to contain the same construct as the ABT salmon) and either age-matched non-GE siblings or pooled control samples from age-matched siblings. Mean plasma T3 concentrations in the GH transgenic Atlantic salmon were statistically different from and lower in the GH transgenic Atlantic salmon compared to non-GE siblings but not when compared to the pooled controls. The highest plasma GH levels correlated with the largest GH transgenic Atlantic salmon while an inverse correlation was shown for mean plasma T3 concentrations.

Mean levels of estradiol, testosterone, 17-ketotestosterone, T3, and T4 were not different in the ABT salmon compared with comparator non-GE salmon (Table 15). The apparent difference in IGF1 in mature diploid ABT salmon compared to sponsor control non-GE salmon was relatively small. Analysis of levels of the IGF1 via the MOE evaluation showed that even the highest measured concentrations would not be biologically significant in the background of the existing systemic and dietary hormonal milieu for the consumer. No differences were observed in levels of growth hormone in edible tissues at the level of quantitation for the analytical method.

Further, even if there were increases in the amounts of these normally occurring substances, Chinook or Atlantic salmon growth hormone, they would not likely effect any biologically meaningful interactions with human growth hormone receptors due to interspecies differences in the ability of these substances to bind to homologous receptors in mammals or to cause physiological changes via such binding. This lack of biological interaction is likely to also be true for mammalian and avian food producing species that could possibly eat animal feed containing ABT salmon (see Food Safety Appendix C).

We conclude that the expression of Chinook salmon growth hormone in ABT salmon does not present a new risk of allergic reaction to salmon allergic individuals and is unlikely to cause allergic cross-reactions. No direct food consumption hazards were identified.

2. Characterization of Indirect Food Consumption Hazards

Indirect food consumption hazards and any indirect food consumption risks are those that might arise as the result of changes that insertion of the AquAdvantage construct at the EO-1α locus of the Atlantic salmon genome could pose.

Based on the risk questions posed in the other steps of the hierarchical review (molecular characterization of the construct, molecular characterization of the construct in the GE animal lineage, phenotypic evaluation), no indirect food hazards were identified.

a. Compositional Analysis

Compositional analyses are a longstanding and well-established approach for assessing the safety of novel foods. Compositional analyses permit an assessment of potential indirect effects that may result from the insertion of an rDNA construct into the genome of an rDNA organism that may impact the safety of foods from that organism. These analyses typically include an assessment of the levels of key nutrients, constituents in the particular food that may have a substantial impact in the overall diet, as well as key toxicants where applicable. Such analyses allow for an assessment of potential nutritional and toxicological risk that may result from changes in significant compositional constituents in the food (CAC, 2008).

An indirect effect of the genetic engineering that resulted in the AquAdvantage Salmon may be a change in the composition of the edible tissues. Such a change may pose a hazard to humans by altering the expected nutritional composition of the food, or it may serve as a signal that an underlying change in the metabolism or physiology of the fish has occurred that may pose a toxicological hazard to humans. Either of these may pose nutritional or toxicological risks via the consumption of edible products from AquAdvantage Salmon.

ABT's approach to address potential indirect toxicity associated with the AquAdvantage Salmon was to evaluate compositional differences between the ABT salmon and non-GE Atlantic salmon. Potential adverse outcomes with respect to consumption of ABT salmon addressed in this section included biologically relevant changes in the proximate, vitamin, mineral, amino acid or fatty acid composition of edible tissues from the GE animal that might suggest toxicological or nutritional concerns compared with levels of these analytes in non-GE Atlantic comparator salmon. A compositional analysis study was provided by ABT.

i. A Single-Blind, Comparator-Controlled, Quantitative Analysis of the Composition of Muscle Skin from Diploid and Triploid Atlantic Salmon (Salmo salar) Modified Transgenically with the AquAdvantage Gene Cassette (opAFP-GHc). Covance Laboratories Inc., Wisconsin. Covance Study Identification 7352-100. Study Report AAS-HFS-001. Report dated 22 January 2003

A total of 144 market-sized (2.0 to 7.5 kg) Atlantic salmon were included in the study: diploid and triploid ABT salmon (referred to as "treated", "TX" by ABT) and non-GE Atlantic salmon from the ABT facility (referred to as "sponsor control", "SC" by ABT); and non-GE diploid and triploid Atlantic salmon from commercial farms in Maine and Canada (referred to as "farm control", "FC" by ABT). TX and SC salmon were bred and reared in ABT facilities; these salmon were not raised in the same tank, but were distributed in different tanks according to their ploidy and the presence/absence of the AquAdvantage construct. TX and SC salmon were fed one of three different diets; MCO, MCAB, or MCA (described subsequently). Husbandry conditions, including diets, of FC salmon were proprietary and therefore not available to the ABT. Salmon were collected at two different times (in October 2001 and June 2002) for the study.

Salmon were screened visually for general health status and traits relevant to commercial marketability, including skin and fin condition, color and markings, and general body morphology. Because of differences in rates of growth to market size, TX and control (SC and FC) fish that were weight-matched may not have been age- matched.

Screening, harvesting, measurements, necropsy and genotype and ploidy analyses were performed by ABT for TX and SC salmon. For FC salmon, screening, harvesting, measurements and necropsy were performed by the salmon farm; ABT performed genotype and ploidy analyses. Blind-coded salmon fillets were frozen and stored at -70°C. Frozen samples were shipped to two different testing laboratories for compositional analysis and hormone analysis.

Tissue samples from a total of 73 salmon were analyzed for proximates, mineral, vitamin, amino acid and fatty acid content. See Part C.2.a.iii, below, for selection criteria. Validation information for each of the assay methodologies was provided. Precision and accuracy results for these analyses are provided. Table 20, below, provides the lower limit of quantitation for the analytes. This study was conducted in compliance with GLPs.

ii. Summary of the Compositional Analysis Results of Study AAS-HFS-001

The arithmetic mean and standard deviation and the maximum and minimum values of compositional analytes (proximates, vitamins, minerals, amino acids and fatty acids) for the farm control (FC), sponsor control (SC) and ABT salmon (TX) are summarized in Tables 21-25 below. Analysis of each group is pooled for gender and ploidy.

iii. Analysis of Study Results

ABT's Analysis

Compositional analysis studies should provide information to allow for the detection of differences among the "test" and comparator groups. ABT initially selected 144 fish for the study, including 61 TX, 54 SC, and 29 FC fish. However, ABT elected to follow the advice of consulting statisticians, thereby selecting only 60 fish for analysis.

ABT's subsequent analysis identified 14 primary sampling units (PSU) based on the availability of TX and SC groups in each "catch" and "demographic trio." Sex, maturity, and ploidy defined a demographic trio, and trios were included in the analysis if at least three fish from each group were available. Only five of the possible eight trios were included in the analysis. Seven fish pairs were identified based on catch, trio, and group. The collected data were analyzed including a pair, group (TX and SC), and the pair-by-group interaction in the statistical model. The pair-by-group- interaction mean square error was used to test for mean differences between the TX and SC group.

We chose not to use this analysis for several reasons, chief among which is our determination that such demographic trios have minimal relevance to food safety concerns for this evaluation. In particular, sex and maturity level were not relevant to the food safety issues confronted by consumers when purchasing food. In addition, we did not concur with ABT's statistical analysis.

Table 20. Lower Limit of Quantitation ($LLOQ_{assay}$) for Analytes for Compositional Analysis

Analyte	$LLOQ_{assay}$
Total (individual) amino acids	0.010 g/100 g (0.010%)
Ash	0.1%
Total carbohydrate	0.10%
Fat by Soxhlet Extraction	0.10%
Fatty acids as triglycerides	0.004%-0.020%*
Folic acid	0.06 μg/g
Free fatty acids by titration	0.01%
Calcium	1.00 mg/100g
Copper	0.0250 mg/100g
Iron	0.100 mg/100g
Magnesium	1.00 mg/100g
Manganese	0.0150 mg/100g
Phosphorus	1.00 mg/100g
Potassium	5.0 mg/100g
Sodium	5.00 mg/100g
Zinc	0.0200 mg/100g
Moisture	0.1%
Niacin	0.3 μg/g
Pantothenic Acid	0.4 μg/g
Protein	0.1%
Selenium	0.030 ppm
Vitamin A	50.0 IU/100g
Vitamin B1	0.01 mg/100g
Vitamin B2	0.2 μg/g
Vitamin B6	0.07 μg/g
Vitamin B12	0.0012 μg/g
Vitamin C	1.0 mg/100g

* The lower limit of quantitation for fatty acids was dependent upon the amount of fat extracted from the sample.

Table 21. Results of Analysis of Proximate Analytes in ABT (TX), Non-GE Sponsor Control (SC) and Non-GE Farm Control (FC) Salmon

Analyte	Group	N	Arithmetic Mean	Standard Deviation	Minimum	Maximum
Carbohydrate	FC	9	0.46	0.357	0.1	1
	SC	22	0.37	0.167	0.1	0.6
	TX	16	0.38	0.335	0.1	1.3
Ash	FC	10	1.13	0.164	0.9	1.4
	SC	33	1.18	0.160	0.8	1.4
	TX	30	1.14	0.218	0.7	1.6
Moisture	FC	10	64.4	2.068	61.1	68
	SC	33	69.3	1.990	64.1	75.2
	TX	30	65.2	3.249	57.4	73.7
Protein	FC	10	18.85	0.610	18.2	19.9
	SC	33	20.16	0.965	15.7	21.4
	TX	30	19.13	1.341	16.3	21.6
Total fat	FC	10	15.17	2.106	11.2	18.9
	SC	33	9.14	1.686	4.5	14.8
	TX	30	14.42	4.123	3.6	24.1

Table 22. Results of Analysis of Vitamins* in ABT (TX), Non-GE Sponsor Control (SC) and Non-GE Farm Control (FC) Salmon

Analyte	Group	N	Arithmetic Mean	Standard Deviation	Minimum	Maximum
Folic acid	FC	10	0.29	0.142	0.15	0.58
	SC	33	0.25	0.092	0.13	0.5
	TX	30	0.22	0.073	0.09	0.41
Niacin	FC	10	88.89	4.375	80.7	96.4
	SC	33	88.66	8.231	63.5	100
	TX	30	97.46	9.164	80.7	118
Pantothenic acid	FC	10	13.40	5.469	5.75	21.6
	SC	33	13.12	2.460	9.09	17.1
	TX	30	11	2.177	6.89	14.8
Vitamin B1	FC	10	0.06	0.014	0.05	0.1
	SC	33	0.08	0.012	0.06	0.11
	TX	30	0.07	0.012	0.04	0.09
Vitamin B12	FC	10	0.03	0.008	0.02	0.05
	SC	33	0.03	0.007	0.02	0.04
	TX	30	0.03	0.008	0.01	0.04
Vitamin B2	FC	10	1.01	0.089	0.86	1.2
	SC	33	1.13	0.143	0.83	1.49
	TX	30	1.08	0.101	0.90	1.28
Vitamin B6[§]	FC	10	6.56	0.593	5.76	7.67
	SC	33	7.20	0.739	4.86	8.72
	TX	30	7.67	0.791	6.50	10.21
Vitamin C	FC	10	2.77	1.069	1.6	4.5
	SC	33	3.98	1.311	1.8	7.5
	TX	30	2.98	0.780	1.6	4.6

*Vitamin A was below the limit of quantitation in all samples and was not included in the evaluation.
§ Vitamin B6 concentrations are reported as the free base form. See Food Safety Appendix E.

Table 23. Results of Analysis of Minerals in ABT (TX), Non-GE Sponsor Control (SC) and Non-GE Farm Control (FC) Salmon

Analyte	Group	N	Arithmetic Mean	Standard Deviation	Minimum	Maximum
Calcium	FC	10	31.49	4.310	25	37.7
	SC	33	30.03	6.260	17.6	43.5
	TX	30	27.57	6.531	16.1	43.4
Copper	FC	10	0.06	0.014	0.04	0.08
	SC	33	0.07	0.014	0.05	0.11
	TX	30	0.08	0.050	0.04	0.33
Iron	FC	10	0.52	0.338	0.29	1.43
	SC	33	0.48	0.082	0.37	0.74
	TX	30	0.52	0.233	0.33	1.65
Magnesium	FC	10	25.56	0.789	24.5	26.8
	SC	30	26.96	1.388	21.9	28.9
	TX	30	24.69	2.265	20.5	29.3
Manganese	FC	10	0.03	0.012	0.02	0.06
	SC	33	0.03	0.045	0.02	0.28
	TX	30	0.03	0.008	0.02	0.06
Phosphorous	FC	10	260.7	3.683	254	267
	SC	33	268.3	13.452	219	285
	TX	30	256.4	17.136	214	291
Potassium	FC	10	375.5	9.606	361	386
	SC	33	393.8	21.760	300	422
	TX	30	368.6	24.795	311	409
Selenium	FC	10	0.20	0.018	0.18	0.23
	SC	33	0.18	0.015	0.14	0.21
	TX	30	0.17	0.011	0.14	0.20
Sodium	FC	10	32.47	2.266	29.2	36.2
	SC	33	35.81	4.322	28.8	47.9
	TX	30	32.53	6.323	25.4	52.6
Zinc	FC	10	0.57	0.096	0.45	0.74
	SC	33	0.52	0.071	0.42	0.73
	TX	30	0.51	0.075	0.39	0.7

Table 24. Results of Analysis of Amino Acids in ABT (TX), Non-GE Sponsor Control (SC) and Non-GE Farm Control (FC) Salmon

Analyte	Group	N	Arithmetic Mean	Standard Deviation	Minimum	Maximum
Alanine	FC	10	1.09	0.044	1.04	1.17
	SC	33	1.17	0.061	0.92	1.27
	TX	30	1.10	0.083	0.96	1.26
Arginine	FC	10	1.06	0.037	1.02	1.13
	SC	33	1.15	0.058	0.90	1.24
	TX	30	1.09	0.075	0.93	1.25
Aspartic acid	FC	10	1.78	0.068	1.7	1.89
	SC	33	1.94	0.099	1.51	2.08
	TX	30	1.82	0.134	1.54	2.08
Cysteine	FC	10	0.21	0.010	0.20	0.23
	SC	33	0.23	0.011	0.19	0.25
	TX	30	0.22	0.014	0.19	0.25
Glutamic acid	FC	10	2.44	0.082	2.33	2.55
	SC	33	2.63	0.127	2.04	2.79
	TX	30	2.44	0.194	2.09	2.82
Glycine	FC	10	0.93	0.044	0.89	1.04
	SC	33	1.02	0.052	0.82	1.08
	TX	30	0.94	0.056	0.84	1.04
Histidine	FC	10	0.51	0.024	0.48	0.55
	SC	33	0.55	0.034	0.42	0.61
	TX	30	0.53	0.036	0.44	0.61
Isoleucine	FC	10	0.85	0.037	0.80	0.91
	SC	33	0.92	0.053	0.70	1.01
	TX	30	0.88	0.059	0.75	0.99
Leucine	FC	10	1.40	0.050	1.34	1.48
	SC	33	1.52	0.077	1.17	1.63
	TX	30	1.42	0.109	1.21	1.64
Lysine	FC	10	1.64	0.054	1.55	1.71
	SC	33	1.77	0.088	1.37	1.89
	TX	30	1.66	0.118	1.42	1.88
Methionine	FC	10	0.54	0.021	0.52	0.58
	SC	33	0.59	0.033	0.48	0.65
	TX	30	0.56	0.039	0.47	0.64
Phenylalanine	FC	10	0.72	0.029	0.69	0.77
	SC	33	0.79	0.040	0.62	0.85
	TX	30	0.74	0.052	0.64	0.85
Proline	FC	10	0.67	0.034	0.62	0.75
	SC	33	0.73	0.039	0.57	0.8
	TX	30	0.68	0.047	0.59	0.77
Serine	FC	10	0.76	0.027	0.73	0.81
	SC	33	0.81	0.055	0.63	0.92
	TX	30	0.76	0.077	0.63	0.89

Table 24. (Continued)

Analyte	Group	N	Arithmetic Mean	Standard Deviation	Minimum	Maximum
Threonine	FC	10	0.76	0.035	0.71	0.82
	SC	33	0.83	0.045	0.64	0.9
	TX	30	0.79	0.060	0.68	0.93
Tryptophan	FC	10	0.17	0.006	0.16	0.18
	SC	33	0.19	0.016	0.13	0.21
	TX	30	0.18	0.014	0.15	0.21
Tyrosine	FC	10	0.62	0.025	0.6	0.67
	SC	33	0.68	0.036	0.53	0.74
	TX	30	0.65	0.049	0.54	0.75
Valine	FC	10	0.99	0.049	0.93	1.08
	SC	33	1.07	0.063	0.81	1.17
	TX	30	1.01	0.072	0.88	1.15

Table 25. Results of Analysis of Free Fatty Acids and Fatty Acids* in ABT (TX), Non-GE Sponsor Control (SC) and Non-GE Farm Control (FC) Salmon

Analyte	Physiological Name1	Group	N	Arithmetic Mean	Standard Deviation	Minimum	Maximum
Arachidic	20:0	FC	10	0.03	0.005	0.02	0.03
		SC	32	0.01	0.005	0.01	0.02
		TX	30	0.03	0.008	0.01	0.04
Arachidonic	20:4 (*n*-6)	FC	10	0.08	0.01	0.07	0.1
		SC	33	0.06	0.016	0.03	0.12
		TX	30	0.09	0.027	0.03	0.17
Docosahexaenoic	22:6 (*n*-3)	FC	10	1.46	0.234	1.06	1.78
		SC	33	0.96	0.186	0.52	1.58
		TX	30	1.42	0.355	0.4	2.26
Docosapentaenoic	22:5 (*n*-3 or 6)	FC	10	0.44	0.073	0.36	0.57
		SC	33	0.27	0.097	0.12	0.66
		TX	30	0.5	0.146	0.18	0.89
Eicosadienoic	20:2 (*n*-6)	FC	10	0.05	0.009	0.03	0.06
		SC	33	0.04	0.01	0.02	0.06
		TX	30	0.06	0.023	0.01	0.1
Eicosapentaenoic	20:5 (*n*-3)	FC	10	1.17	0.199	0.86	1.44
		SC	33	0.59	0.196	0.29	1.37
		TX	30	1.1	0.346	0.26	2.07
Eicosatrienoic	20:3 (*n*-3)	FC	9	0.02	0.006	0.01	0.03
		SC	29	0.01	0.004	0.01	0.02
		TX	27	0.03	0.011	0.01	0.04

Analyte	Physiological Name1	Group	N	Arithmetic Mean	Standard Deviation	Minimum	Maximum
Eicosenoic	20:1 (*n*-9)	FC	10	0.91	0.106	0.79	1.16
		SC	33	0.46	0.114	0.22	0.64
		TX	30	0.53	0.176	0.18	0.77
Free fatty acids	Variable	FC	9	0.04	0.026	0.01	0.09
		SC	33	0.07	0.028	0.03	0.13
		TX	28	0.09	0.033	0.03	0.17
Gamma linolenic	18:3 (*n*-6)	FC	10	0.03	0.005	0.02	0.04
		SC	33	0.02	0.004	0.01	0.03
		TX	30	0.03	0.007	0.01	0.04
Heptadecanoic	17:0	FC	10	0.04	0.007	0.02	0.04
		SC	33	0.02	0.006	0.01	0.04
		TX	30	0.04	0.011	0.01	0.06
Linoleic	18:2 (*n*-6)	FC	10	0.67	0.105	0.43	0.78
		SC	33	0.51	0.097	0.28	0.68
		TX	30	0.74	0.311	0.14	1.2
Linolenic	18:3 (*n*-3)	FC	10	0.18	0.049	0.12	0.24
		SC	33	0.13	0.039	0.07	0.21
		TX	30	0.23	0.128	0.025	0.42
Myristic	14:0	FC	10	0.75	0.111	0.51	0.92
		SC	33	0.4	0.091	0.19	0.74
		TX	30	0.66	0.196	0.15	1.18
Oleic	18:1 (*n*-9)	FC	10	2.88	0.437	2.2	3.68
		SC	33	2.01	0.328	1.11	2.69
		TX	30	3.3	1.085	0.74	4.98
Palmitic	16:0	FC	10	1.91	0.333	1.17	2.21
		SC	33	1.07	0.262	0.48	2.05
		TX	30	1.79	0.549	0.41	3.39
Palmitoleic	16:1 (*n*-7)	FC	10	0.98	0.138	0.75	1.21
		SC	33	0.56	0.137	0.26	1.06
		TX	30	0.89	0.265	0.23	1.7
Pentadecanoic	15:0	FC	10	0.05	0.008	0.03	0.06
		SC	33	0.03	0.007	0.01	0.05
		TX	30	0.04	0.012	0.01	0.07
Stearic	18:0	FC	10	0.39	0.067	0.25	0.46
		SC	33	0.24	0.061	0.11	0.48
		TX	30	0.42	0.131	0.11	0.82

* Caprylic, capric, pentadecenoic, heptadecenoic, erucic and lignoceric acids were below the limit ofquantitation in all samples; lauric, myristoleic, and behenic acids were below the limit of quantitation in all but one sample. These analytes were not included in our evaluation.

[1] Physiological name is biochemical nomenclature for both lipid number, degree of saturation, and omega number.

CVM's Analysis

(a) General Approach

Characteristics of individual fish, e.g., sex or season of harvest (time of catch), may have an impact on their composition. The comparisons of interest are between TX, SC, and FC salmon with consideration of ploidy. If, in general, the relative differences among TX, SC, and FC salmon are the same for both ploidies, then ploidy is not a consideration and comparisons among groups can be made ignoring ploidy. Variability among fish within groups is considered when making the comparisons and inclusion of fish with different characteristics broadens the inference.

Although it may be important to try to identify toxicologically or nutritionally significant compositional differences between TX and SC salmon, we believe it is equally important to identify such differences between TX salmon and salmon normally consumed by humans, such as FC salmon. Considering all these factors, we conclude that comparing the composition of TX salmon to either or both SC and FC control salmon groups is appropriate for determining whether or not TX salmon have important compositional differences from biologically relevant comparator salmon.

Initially, a heuristic evaluation of the data was performed. We compared the arithmetic means of values for each analyte derived from samples from TX fish to the respective means from the SC and FC groups. If the arithmetic mean from the TX fish were equal to or between the arithmetic means of SC and FC groups, i.e., SC $\leq$ TX $\leq$ FC, the results for TX salmon were considered to be similar to "control salmon" results. If the arithmetic mean for the TX salmon fell outside the range of the either the SC or FC group, the minimum and maximum values (extreme values) for the TX salmon were compared to the range of values from the SC and FC salmon. If these extreme values from the TX salmon did not fall outside the range of values from the SC and FC salmon, the results for TX salmon were considered to be similar to "control salmon" results. If TX salmon were not considered similar to "control salmon", individual values were compared and if the individual values for the TX salmon were not more than 10% beyond the range of values for the individual "control salmon," the values for the TX salmon were considered to be within normal biological variability and thus similar to the "control salmon."

Following this heuristic evaluation, if we were unable to consider the TX salmon similar to "control salmon," a statistical analysis was performed. The statistical analysis took into consideration the variability among the fish in each group to test for differences in means. Selected data were analyzed using analysis of variance (ANOVA) with group (FC, SC, TX), ploidy (diploid, triploid), and the group-by- ploidy interaction included in the model as fixed effects. If the group-by-ploidy interaction was considered significant, this indicated that generally the mean results among the groups differed in some way. In this case, the TX group mean was compared to the FC and SC group means separately within ploidy. If the group-by-ploidy interaction was not considered significant, and the group effect was considered significant, the TX group mean was compared to the FC and SC group means without regard to ploidy. Note that the analysis results were interpreted with the understanding that the estimated p-value may be under-estimated because comparisons are generated after the data were examined. However, for exploratory analyses, this is an acceptable strategy. Results of the statistical analyses are provided in Food Safety Appendix D.

(b) Results of CVM Analysis

(i) Analysis of Results of Proximates, Vitamins, Minerals and Amino Acids Based on the comparison of arithmetic means and extreme values, the following analytes from TX salmon are considered to be similar to those for SC and FC (comparator) salmon:

- **proximates** - carbohydrate, ash, moisture, protein and total fat;
- **vitamins** – pantothenic acid, vitamins B1, B12, B2 and C;
- **minerals** – potassium, selenium and sodium; and
- **amino acids** – alanine, arginine, aspartic acid, cysteine, glutamic acid, glycine, histidine, soleucine, leucine, lysine, methionine, phenylalanine, proline, threonine, tryptophan, tyrosine and valine.

or the remaining analytes (**vitamins** - folic acid, niacin and vitamin B6; **minerals** - calcium, copper, iron, magnesium, manganese, phosphorous and zinc; and **amino acid** – serine) that did not meet the criteria identified above, the individual values from the TX salmon were again compared to the individual values from the two comparator samples using a 10% exceedance range.

In all of these 11 analytes, at least one value exceeded the range of values for control salmon by at least 10%; five analytes had one value from TX salmon that exceeded the non-GE range by more than 10% (i.e., calcium, copper, iron, manganese and serine). The remaining six analytes had between 2 and 8 values from TX salmon that exceeded the comparator range by 10% or more (i.e., folic acid, niacin, vitamin B6, magnesium, phosphorous and zinc).

From the statistical analyses, analytes for which no statistical difference was detected when ploidy was considered as a variable for each group included calcium, copper, manganese, serine, vitamin B6, and zinc. When results from analytes from all TX salmon were compared to results from all the comparator groups, SC and FC salmon, four were not statistically significantly different from either SC or FC salmon (calcium, copper, manganese and zinc), two, potassium and serine, were statistically significantly different from SC salmon but not from FC salmon and one (vitamin B6) was statistically significantly different from both SC and FC salmon (see **(ii)**, below).

From the statistical analyses, analytes for which the results were affected by ploidy were iron, phosphorous, folic acid, magnesium and niacin. When these analyte results for *diploid* TX salmon were compared to results for diploid SC and FC salmon, three were not statistically significantly different from either group of control salmon (iron, magnesium, and phosphorous), and two were statistically significantly different from both SC and FC salmon (folic acid and niacin). When these analyte results for *triploid* salmon were compared to results for triploid SC and FC salmon, two were not statistically significantly different from either group of control salmon (folic acid and iron), three were statistically significantly different from SC salmon but not from FC salmon (niacin, magnesium, and phosphorous).

Conclusions for proximate, vitamin, mineral, and amino acid analytes.

Of the proximate, vitamin, mineral and amino acid analytes in this study, only three analytes were present at levels in TX salmon that were statistically significantly different from levels in both control salmon (SC and FC): *vitamin B6* (when diploid and triploid salmon were considered together for TX, SC and FC salmon), *folic acid* (when diploid salmon were compared in TX, SC and FC salmon) and *niacin* (when triploid salmon were compared in TX, SC and FC salmon). Based on all previous criteria including statistical analysis, we conclude that the levels of all proximate, vitamin, mineral and amino acid analytes in TX salmon except vitamin B6 are similar to levels in one or more appropriate groups of control salmon.

(ii) Vitamin B6

The levels of Vitamin B6 in the diploid ABT salmon were statistically different from both the comparator groups. In order to determine whether there was any biological relevance to this statistical difference, we performed a Margin of Exposure assessment using the same assumptions as for the analysis of IGF1 (Table 19: MOE for IGF1; Table 27: MOE summarizes the analysis for Vitamin B6).

Margin of Exposure for Vitamin B6

Vitamin B6 concentrations were slightly elevated in the TX salmon samples: this elevation was determined to be statistically significant. Prior to conducting any further analyses to determine whether this statistically significant elevation had any biological implications, we performed two assessments: (1) visual inspection of the data to determine the distribution of values among the diploid and triploid TX samples to determine the source of the elevation, and (2) a margin of exposure analysis (MOE) on the most elevated level to ascertain that even if all of the fish consumed contained that highest level, whether exposures would pose a hazard.

We used the most extreme value as the intrinsic exposure level in the MOE assessment. The same assumptions for consumption values were used in the MOE for vitamin B6 as were used for the MOE analysis of IGF1 (Table 18; MOE IGF1) (Table 16; MOE vitamin B6). The margin of exposure between B6 in non-GE Atlantic salmon and diploid ABT salmon are presented in Table 27.

Comparisons to other sources of B6 from finfish and other common protein sources are also appropriate to put this analysis into context. This information is summarized in Table 26.

Results

The recommended maximum level of daily nutrient intake (UL) for vitamin B6 is **100 mg/day** for healthy male adults aged 13 and above (IOM, 2001; WHO-FAO, 2004). Healthy, non-pregnant women aged 19 and above are also advised to consume no more than **100 mg/day** of vitamin B6 (IOM, 2001; WHO-FAO, 2004).

We calculated that the upper bound dietary consumption level of vitamin B6 from sponsor control mature diploid non-GE salmon, and mature diploid ABT salmon to be **1.74 mg/day** and **2.04 mg/day** respectively.

The MOE for dietary consumption of vitamin B6 was therefore reported as the index number of the maximum level of daily intake divided by the upper bound dietary consumption level [i.e., (100 mg/ day UL)/(2.12 mg/day)]. For sponsor control non- GE salmon, the MOE yielded **57.5 fold less than the maximum** allowable consumption level. For mature diploid ABT salmon, the MOE was calculated as **49.0 fold less than the maximum** allowable consumption level.

Table 26. Vitamin B6 Concentration Found in Commonly Eaten Protein Sources[1] (Reported as mg/100 g Tissue Mass)

Source	Mean	Minimum	Maximum
Sponsor-Provided Data			
ABT Salmon*	0.77	0.65	1.02
Non-GE Sponsor Control Salmon*	0.72	0.49	0.87
USDA Nutrient Database			
Atlantic Salmon (farm and wild)	0.73	0.64	0.82
Chinook Salmon (farm and wild)	0.43	0.4	0.46
Chum Salmon (farm and wild)	0.55	0.38	0.94
Coho Salmon (farm and wild)	0.6	0.55	0.66
Sockeye Salmon (farm and wild)	0.23	0.19	0.28
Rainbow Trout (farm and wild)	0.51	0.41	0.62
Tuna	0.81	0.46	1.04
Flatfish	0.28	0.16	0.42
Beef	0.46	0.36	0.56
Pork	0.49	0.21	0.75
Poultry	0.54	0.25	1.47
Milk	0.05	0.04	0.05

[1] U.S. Department of Agriculture, Agricultural Research Service (2009) USDA National Nutrient Database for Standard Reference, Release 22. Nutrient Data Laboratory Home Page, http://www.ars.usda.gov/nutrientdata

*ABT and Sponsor Control Salmon Vitamin B6 concentrations calculated and reported as free base form. See Food Safety Appendix E.

Conclusions

The statistically significant difference in mean vitamin B6 levels was investigated using a margin of exposure assessment. Even if the highest observed level of vitamin B6 observed in the diploid ABT salmon were to be found in all ABT salmon, the margin of exposure assessment indicated that it would still be well within the upper bound recommended daily intake for vitamin B6. We therefore find that there is no food consumption hazard due to vitamin B6.

(iii)Fatty Acids and Free Fatty Acids

Comparison of TX, SC and FC means and ranges

Based on the comparison of arithmetic means and extreme values, the following fatty acids for TX salmon were considered to be present at similar levels in TX and control salmon: arachidic, docosahexaenoic, eicosanoic, eicosapentaenoic, gamma linolenic, heptadecanoic, myristic, palmitic, palmitoleic and pentadecanoic fatty acids.

For the remaining fatty acids not considered to be present at similar levels in TX and control salmon according to the criteria described in Part (b)(i) above (for proximates, vitamins, minerals, and amino acids), the individual values for the TX salmon are again compared to the individual values from the two comparator samples using a 10% exceedance range.

Table 27. Margin of Exposure Estimates of Vitamin B6 in ABT Salmon

Description	Calculation	Value
Daily non-tuna consumption for the 95th percentile eater[1]		300 g/day
Assuming all consumed non-tuna finfish are salmon, consumption corrected for fraction of Atlantic salmon	(2/3) * 300 g/day	200 g/day
Plausible upper bound of Vitamin B6 concentration observed at maximum level in sponsor control mature diploid Atlantic salmon (μg of Vitamin B6/g total weight)		8.7 μg/g
Plausible upper bound of Vitamin B6 concentration observed at maximum level in sponsor control mature diploid ABT salmon (μg of Vitamin B6/g total weight)		10.2 μg/g
Daily consumption of Vitamin B6 from non-GE Atlantic salmon	200 g/day * 8.7 μg/g	1.74 mg/day
Daily consumption of Vitamin B6 from ABTsalmon	200 g/day * 10.2 μg/g	2.04 mg/day
Recommended maximum level of dailyVitamin B6 intake[2,3]		100 mg/day
Margin of Exposure (MOE)		
MOE for mature diploid (non-GE) Atlantic salmon	(100 mg/day)/(1.74 mg/day)	57.5 fold
MOE for mature diploid ABT salmon	(100 mg/day)/(2.04 mg/day)	49.0 fold

[1] US Department of Commerce (USDOC), Technology Administration, NTIS, Springfield, VA 22161 (2002) Foods Commonly Eaten in the United States. Quantities Consumed per Eating Occasion and In A Day, 1994-1996. PB2005110468.

[2] Institute of Medicine (IOM) of the National Academies. Food and Nutrition Board (2001) Dietary Reference Intakes: The National Academies Press. Washington, DC.

[3] World Health Organization (WHO) and the Food and Agriculture Organization of the United Nations (FAO) (2004) Vitamin and Mineral Requirements in Human Nutrition. 175-179. Geneva, Switzerland.

The values for one fatty acid, eicosatrienoic acid, were entirely within the range of controls (or within 10%); these were considered to be similar to results in control salmon by the criteria described in Section b.*i*. However, the remaining eight fatty acid analytes in TX salmon appeared to be dissimilar to those in control salmon: the levels of three analytes (the free fatty acids, arachidonic acid, and docosapentaenoic acid) were more than 10% beyond the range of control values, and for the remaining 5 fatty acid analytes (stearic acid, oleic acid, linoleic acid, linolenic acid and eicosadienoic acid), 8-16 measurements in the TX group were more than 10% beyond the range of controls values. Eicosadienoic, linoleic, linolenic, oleic, arachidic, docosahexaenoic, docosapentaenoic, eicosopentaenoic, palmitoleic, palmitic, and stearic acids and free and total fatty acids were statistically analyzed.

From the statistical analyses for eicosadienoic, linoleic, linolenic, oleic and palmitic fatty acid analytes, the group-by-ploidy interaction was not considered statistically significant but the group effect was. Mean comparison were made among groups for these fatty acid analytes. For eicosadienoic, linoleic, linolenic and oleic fatty acids, the TX mean was statistically significantly different from the SC mean. For palmitic acid, no statistically significant differences were found between TX mean and the FC or SC mean.

From the statistical analyses for arachidic, docosahexanoic, docosapentaenoic, eicosapentaenoic, palmitoleic, palmitic and stearic and total and free fatty acid analytes, the group-by-ploidy interaction was considered significant and mean comparisons were made within ploidy. For diploid salmon, statistically significant differences were found between the TX mean and the SC mean for docosahexanoic, docosapentaenoic, eicosapentaenoic and stearic fatty acids and total fatty acids, and between the TX mean and both the SC and FC means for arachidic, palmitic and palmitoleic fatty acids. For diploid salmon, no statistically significant difference was found between the TX mean and either SC or FC mean free fatty acids. For triploid salmon, statistically significant differences were found between the TX mean and the SC mean for arachidic, docosahexanoic, docosapentaenoic, eicosapentaenoic, palmitic and palmitoleic fatty acids and total fatty acids and between the TX mean and both the SC and FC means for stearic acid and free fatty acids.

Of the fatty acid analytes in this study, only four were present at levels in TX salmon that were statically significantly different from levels in both SC and FC control salmon – fatty acids arachidic, palmitic and palmitoleic (when diploid salmon were compared in TX, SC and FC salmon) and stearic acid (when triploid salmon were compared in TX, SC and FC salmon). Thus, based on all previous criteria, including statistical analysis, we conclude that the levels of all fatty acid analytes in TX salmon are similar to levels in one or more appropriate groups of control salmon.

Comparison of dietary fat, total fat levels and levels of free fatty acids

For this study, we noted that the variation in total fat levels within each group of salmon was fairly broad, with arithmetic mean values of 14.4%, 15.2% and 9.1% and standard deviations of 4.12, 2.11 and 1.69 for TX, FC and SC salmon, respectively.

Although the arithmetic mean total fat content for TX salmon was similar to that of the FC salmon – which was our basis for concluding that total fat levels in TX salmon were similar to those for control salmon - it was higher than the total fat content of the SC salmon. Nonetheless, we noted that the total fat level arithmetic means for all of the salmon in this study were within the 7% - 19% range for total lipids as reported in scientific literature for wild and farmed Atlantic salmon (Hamilton et al., 2005; Torstensen et al., 2005).

The amount of total fatty acids in fish is generally directly proportional to total lipid deposition (Ikonomou et al., 2007); this relationship is also observed in Figure 4 generated using ABT's data. Thus, because the level of total lipids in TX salmon is higher than in SC salmon, it is reasonable to expect that the levels of some fatty acids in TX salmon will be proportionately higher as well. This pattern is particularly evident for the following fatty acids: eicosadienoic, linoleic, linolenic, oleic, docosahexanoic, docosapentaenoic, eicosapentaenoic and stearic and total fatty acids; for these fatty acid analytes, levels in TX salmon are similar to levels in FC salmon but are higher than levels in SC salmon.

The effect of dietary lipid levels and their origin (plant versus animal) in the diet of fish on the fatty acid composition of fish has been extensively studied (Friesen et al., 2008;

Kennedy et al., 2005; Polvi and Ackman, 1992; Torstensen et al., 2005). The composition and amount of total fat and fatty acids in salmon vary from species to species and is, in general, a function of the overall composition of their feed and, in particular, on the amount of dietary fat and fatty acids in the diet. According to the information provided by ABT, most of the ABT and sponsor control salmon participating in the study were fed the Moore-Clark Orion (MCO) commercial grower diet. Smaller groups of the fish were fed the Moore-Clark Atlantic (MCA) and the Moore-Clark Atlantic Broodstock (MCAB) commercial diets. The composition of these diets was provided. All three commercial diets can be considered "high energy" diets due to their high protein content (MCO 37%, MCA 43% and MCAB 46%) and their high crude fat content (MCO 36%, MCA 32%, and MCAB 25%). The other control group used in this study, farm-raised fish (FC), was purchased from a commercial Atlantic salmon farm facility. The information regarding the feed formulation used at that facility is considered proprietary and was not provided, limiting the extent to which comparisons could be made. Generally, fish on a "high energy" diet exhibit elevated total lipid levels. ABT compared the protein and fat content of two of the diets fed to TX and SC salmon during the three months before they were killed with the fat levels in the salmon. (No TX salmon were fed the third diet for the three months before they were killed.) TX fish fed both diets appear to have higher percentages of total fat than SC salmon, even when the diets had identical fat content. As noted previously, all three diet formulations used for TX and SC salmon contained plant based oils as their main source of lipids. Because TX and SC fish were fed similar diets during the three months prior to collection, observed differences in their total fat content cannot be attributed to different levels of energy, protein, or total fat in their diets or to differences in sources of dietary fats in the diets (e.g., animal versus plant).

Because Atlantic salmon are an important source of lipids and, more specifically, of polyunsaturated omega-3 and omega-6 fatty acids in our diet, consumers may be concerned that consuming ABT salmon may result in lower intakes of omega-3 and omega-6 fatty acids or a change in the ratio of these fatty acids. Examination of omega-3/omega-6 ratios showed that they were virtually identical across the TX, SC, and FC groups and are similar to the ratios found in scientific literature for farmed Atlantic salmon. These data, along with data from published literature, are presented in Table 28.

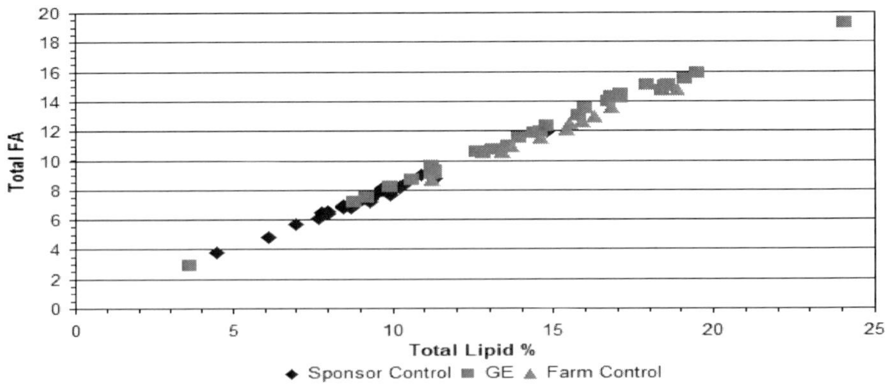

Relationship between total fatty acids and total lipids as a percentage of body weight (Blue Diamonds: Sponsor Controls; Pink Squares: ABT salmon; Green Triangles: Farm Controls).

Figure 4. Total Fatty Acids vs. Total Lipid %.

Table 28. Omega-3 and Omega-6 Fatty Acid Levels in ABT Salmon and Farmed Atlantic Salmon

Fatty Acid	Degree of saturation	Atlantic salmon							
		Aqua Bounty			Scientific literature				
		Farm raised	Sponsor control	ABT	Farmed[1]	Farmed[2]	Farmed[3]	Wild caught[4]	Farmed[4]
		Means (% of the wet weight)							
Oleic	18:1	2.88	2.011	3.299		0.465	1.05		
Linoleic¶	18:2	0.668	0.507	0.743	0.303	0.162	0.194	0.067	0.65
α-Linolenic §	18:3	0.178	0.131	0.232	0.066	0.031	0.103	0.05	0.181
γ-Linolenic ¶	18:3	0.03	0.019	0.027				0.003	0.014
Arachidonic¶	20:4	0.084	0.055	0.092	0.037	0.025	0.029	0.03	0.091
Eicosenoic	20:1	0.913	0.455	0.534		0.144	0.275		
Eicosadienoic¶	20:2	0.053	0.039	0.059			0.029	0.017	0.063
Eicosatrienoic§	20:3	0.021	0.012	0.024				0.009	0.024
Eicosapentaenoic§	20:5	1.174	0.593	1.095	0.324	0.225	0.326	0.414	1.08
Docosapentaenoic§	22:5	0.436	0.266	0.5				0.12	0.519
Docosahexaenoic§	22:6	1.46	0.961	1.422	0.623	0.568	0.932	0.629	1.57
ω-3/ω-6 ratio		3.9	3.2	3.6				10.	4.

§ Omega - 3 fatty acids.

¶ Omega - 6 fatty acids.

[1] Blanchet C, M. Lucas, P. Julien, R. Morin, S. Gingras, and E. Dewailly (2005) Fatty acid composition of wild and farmed Atlantic Salmon (*S. salar*) and Rainbow trout (*O. mykiss*). *Lipids* 40:529.

[2] Kennedy S.R., P.J. Campbell, A. Porter, and D.R. Tocher (2005) Influence of dietary conjugated linoleic acid on lipid and fatty acid composition in liver and flesh of Atlantic salmon (*S.salar*). *Comp. Biochem. Phys.*, Part B,141:168.

[3] Bell J.G., R.J. Henderson, D.R. Tocher, F. McGhee, J.R. Dick, A. Porter, R.P. Smullen, and J.R. Sargent (2002) Substituting fish oil with crude palm oil in the diet of Atlantic Salmon (*S. salar*) affects muscle fatty acid composition and hepatic fatty acid metabolism. *J Nutr.*132, 222.

[4] Hamilton M.C., R.A. Hites, S.J. Schwager, J.A. Foran, B.A. Knuth, and D.O. Carpenter (2005) Lipid composition and contaminants in farmed and wild salmon. *Environ. Sci. Technol.* 39:8622.

Conclusions for Fatty Acids

Based on the data and information evaluated, we conclude the following:

- The levels of any individual fatty acid in ABT salmon are similar to those of the comparators;
- Total fat content for ABT salmon is similar to those for comparator salmon and within the 7-19% range for total lipids as reported for wild and farmed Atlantic salmon;
- ABT salmon offer a balanced diet of omega-3 and omega-6 fatty acids, similar in quantity and ratio to that provided by the FC controls and by farmed Atlantic salmon currently consumed; and

- ABT salmon are not materially different from other Atlantic salmon with respect to omega-3 and omega-6 fatty acid levels and the ratio of omega-3 to omega-6 fatty acids.

Composition Conclusions

Based on all previous criteria including statistical analyses, we conclude that levels of all analytes in ABT salmon are similar to levels in appropriate comparator salmon (e.g., either the sponsor controls, farmed salmon, literature reports, or some combination of the three).

We conclude that any differences observed for analytes are the result of normal biological variation, and are highly unlikely to be associated with toxicological or nutritional hazards to humans consuming ABT salmon.

The statistically significant difference in mean vitamin B6 levels was investigated using a margin of exposure assessment. Even if the highest observed level of vitamin B6 observed in the diploid ABT salmon were to be found in all ABT salmon, the margin of exposure assessment indicates that it would still be well within the upper bound recommended daily intake for vitamin B6.

For fatty acid analytes, values found in the ABT salmon are consistently more similar to the farmed control values than to the sponsor control values, and are proportional to total fat levels in these three groups of salmon; these differences do not appear to be due to differences in fat content or protein source of the ABT and sponsor control salmon diets.

ABT salmon offer a balanced diet of omega-3 and omega-6 fatty acids, similar to that provided by the FC controls and by farmed Atlantic salmon.

Finally, we conclude that ABT salmon are not materially different from other Atlantic salmon with respect to omega-3 and omega-6 fatty acid levels and the ratio of omega-3 to omega-6 fatty acids.

b. Endogenous Allergenicity

i. Context

The Food Allergen Labeling Consumer Protection Act of 2004 (P.L. 282) (FALCPA, 2004) identifies eight major foods or food groups that are allergenic: milk, eggs, fish, crustacean shellfish, tree nuts, peanuts, wheat, and soybeans. These eight foods are believed to account for 90 percent of food allergies and most serious reactions to foods in the U.S. (FALCPA, 2004; Hefle et al., 1996). Each of these eight major allergenic foods contains multiple allergenic proteins, many of which have not been fully characterized (Gendel, 1998).

Food allergies affect more than 1-2% but less than 10% of the U.S. population (NIAID, 2010). Although there are numerous scientific publications on food allergy, there are many uncertainties with respect to diagnosis, best practices for management and prevention, mechanisms of sensitization, and allergenic thresholds that will elicit responses from sensitive individuals (CFSAN, 2006; Chafen et al., 2010).

There are a great number of uncertainties when attempting to assess potential changes in the levels of allergens in commonly allergenic foods, including salmon. We are unaware of any data that may exist on the natural variation in the levels of endogenous allergens in salmon or other finfish that are currently consumed in the U.S. Tools to assess endogenous allergen levels are limited. Human sera containing specific IgE are often used for this purpose; however, the utility of these studies is limited because there is typically little or no information available regarding the allergic history of the donors. This is important because the presence of specific IgE in sera does not necessarily correlate with a clinically relevant food allergy (Chafen et al., 2010); therefore, the allergic history of the donor should be taken into account when interpreting data from such a study. In addition, because of the relatively small numbers of individuals represented in such assays, these *in vitro* studies may not reflect the responses of a general population. Finally, there is no consensus in the scientific and medical communities regarding the magnitude of increase in endogenous allergens in an allergenic food that would present an additional risk to public health (Goodman et al., 2008), especially considering that individuals who are allergic to a particular food would likely avoid that food.

Because finfish are one of the major eight allergenic foods in the United States (FALCPA, 2004; Hefle et al., 1996; Sampson, 2004) one potential indirect hazard that may result from the insertion of the AquAdvantage construct at the α- locus is an alteration in the endogenous levels of allergens in ABT salmon due to insertional mutagenesis. In particular, the question was asked whether the edible tissue from GE salmon is more allergenic than the non-GE comparator. This question was evaluated in the ABT submission discussed below.

ii. Sponsor Study: A Comparator-Controlled Immunochemical Study of the Allergenic Potency of Muscle-Skin from Diploid and Triploid Atlantic Salmon (Salmo salar) Modified Transgenically with the AquAdvantage Gene Construct opAFP-GHc2). Testing Facility: IBT Reference Laboratory. Kansas. Study Report AAS-HFS-003. Report dated 22 March 2006

The purpose of this study was to examine potential quantitative and qualitative changes in allergens in salmon muscle and skin from market-size, diploid and triploid ABT salmon vs. non-GE Atlantic salmon. This study was conducted in compliance with GLPs.

Overall study design:

Market-sized (2.0 to 7.5 kg) diploid and triploid ABT salmon (treated, TX_D and TX_T) and non-GE diploid Atlantic salmon (sponsor control, SC_D) were included in the study. From the available pools of each type fish, six fish were selected non- systematically by net capture for a total of 18 fish. Sex and maturity were not considered for selection, therefore distribution was not uniform within or between groups.

Salmon were screened visually for general health status and traits relevant to commercial marketability, including skin and fin condition, color and markings, and general body morphology. Because of differences in rates of growth to market size, ABT and control fish may not have been age-matched.

Harvest, measurements, necropsy, genotype, and ploidy determinations were performed by ABT. Blind-coded salmon fillets packed on dry ice were shipped to a testing laboratory, which homogenized the samples under liquid nitrogen. A representative subsample of each frozen salmon-fillet homogenate (FSFH) was shipped on dry ice to IBT for testing.

ABT subsequently unblinded the identities of all 18 samples to facilitate use of control FSFH in further analyses. IBT performed aqueous extractions of a subsample of each FSFH, and extracts were stored at -70°C. Separate aqueous extracts from the same FSFH samples were used in the allergen potency and allergen identity assays due to insufficient quantity of extracts. IBT also performed total protein determination of extracts as well as allergen potency and identity assays.

Total protein concentration from salmon skin-muscle extracts was determined using the Micro BCA™ Protein Assay Kit (Pierce Chemical Company), in accordance with GLPs. Validation information for assay methodology was provided and the limit of detection of the assay was < 2 mg/ml; the lower limit of quantitation (LLOQ) was 3.1 mg/ml. All extracts were normalized to 2 mg/ml prior to further analysis.

c. Fluorescent Enzymatic Immunoassay (FEIA)

IBT developed an inhibition assay to determine relative allergenic potency (RP) of FSFH extracts based on the ImmunoCAP system by Pharmacia Diagnostics AB (now Phadia AB). The ImmunoCAP system is a commercial reagent and equipment system used for clinical diagnostic testing of human sera for specific IgE. Briefly, solid phase bound allergen standard is allowed to react with IgE antibodies in a serum sample. IgE antibodies bound to the allergen standard are detected by β-galactosidase labeled anti-IgE antibody. β-galactosidase catalyzes the hydrolysis of a substrate yielding fluorescence, providing a quantitative measure of IgE binding to the allergen standard.

Instrumentation, methods, reagents and salmon-allergen standard used in the FEIA were developed and validated for the commercial use of the ImmunoCAP system by Pharmacia Diagnostics AB. The assay was conducted in accordance with GLPs.

IBT used this FEIA as the basis for the development of an inhibition assay to determine the allergenic potency of muscle-skin extracts from ABT salmon compared with extracts from sponsor control non-GE salmon. Briefly, soluble salmon allergen in FSFH extracts is used to inhibit binding of highly salmon-reactive IgE pooled sera to the solid phase bound salmon standard f41 which is derived from the muscle of Atlantic salmon. This assay provides a quantitative determination of inhibition of salmon-specific IgE binding which is then used to calculate the potency of salmon allergen in muscle-skin from GE salmon relative to that in a control extract, comprised of equal volumes of all six sponsor control non-GE FSFH extracts.

Individual human sera with salmon-specific IgE of greater than or equal to Class 3 by ImmunoCAP scoring guide (greater than or equal to 3.5 kU/l, with individual sera ranging from 4.8 – 98.60 kU/l) were obtained commercially and pooled. IBT determined binding characteristics of salmon-specific IgE pool by ImmunoCAP for use in the FEIA inhibition assay. Individual human sera negative for salmon-specific IgE (less than 0.10 kU/l) were obtained from IBT's sera bank and pooled.

Each of the 18 individual FSFH extracts was run in the FEIA inhibition assay six times. Validation information for assay methodology was provided.

Percent inhibition was calculated and plotted against the log of the reciprocal sample dilution. These data were used to generate an inhibition curve, from which the allergenic potency in U/ml at 40, 50, and 60 % inhibition was calculated for each FSFH sample and pooled FSFH control. Relative potency was estimated using the percent inhibition of pooled FSFH control.

IBT normalized the mean RP values using the mean RP value for the sponsor control fish. The resulting normalized RP values for sponsor control, diploid GE, and triploid GE fish were 1.00, 1.52 and 1.20, respectively. ABT set its acceptability criteria for RP based on FDA's Center for Biologics Evaluation and Review's (CBER) Guidance for Reviewers: Potency Limits for Standardized Dust Mite and Grass Allergen Vaccines: A Revised Protocol

(November 2000) (CBER's Allergen Vaccine Guidance) (Available at http://www.fda.gov/BiologicsBloodVaccines/GuidanceComplianceRegulatoryInformation/Guidances/Allergenics/ucm071931.htm). The release limits established by CBER for standardized dust mite and grass allergen vaccines is 0.5-2.0 RP. Using the normalized RP values and CBER's lot release criteria, ABT concluded that both diploid and triploid GE fish fall within the bounds of an equivalent response vs. control fish.

Evaluation

Our evaluation of the *overall study design* indicated the following notable concerns:

1. The number of samples per group was limited. Fish were included irrespective of sex or maturity so that these were not distributed uniformly within or between the different groups. Because fish were included irrespective of sex or maturity, results may not necessarily be representative of the AquAdvantage Salmon that will be marketed for consumption.
2. Farm-raised salmon were not included as a control. Because farm-raised salmon were selected for rapid growth, inclusion of this group would have provided a control for potential effects related to a rapid growth phenotype. In addition, farmed salmon could have provided additional information regarding the natural variability in the levels of endogenous allergens in salmon currently consumed in the U.S.

In addition, our evaluation of the overall study design indicated the following notable deviations from the study protocol:

The testing laboratory's original principal investigator (PI) departed while analyses were on-going. The original PI was replaced by a subsequent PI. This deviation had no impact on the study.

Study protocol stated that, "*FSFH subsamples deriving from homogenization of the blind-coded left fillets would be sent [] for immunochemical analysis extracts identified only by the UFID [universal fish identification number] originally provided [] by [ABT].*" ABT unblinded the identities of all 18 samples to facilitate use of control FSFH in further analyses by the testing laboratory. Blinding the identities of the samples could have provided some bias control in the outcomes of both fluorescent enzymatic immunoassay (FEIA) inhibition and Western blot analyses.

Our evaluation of the FEIA study design indicated the following notable deviations from the study protocol:

According to the study protocol, human sera with salmon-specific IgE were to have been selected from the testing laboratory's sera bank and pooled. Initial studies including validation were performed using this source of pooled sera. Limitations in the quantity of original salmon-specific IgE pool necessitated the purchase of commercially available sera to complete the inhibition assays on FSFH extracts. We note that although two sets of extracts and salmon-specific IgE sera pools were used for the FEIA, the initial sets were used to establish assay conditions. All six runs of the FEIA were performed using the second set of extracts and the second sera pool. This point was clarified in conversation between CVM and the testing laboratory. These deviations had no impact on outcome of the study.

Study protocol indicates that the relative potency of FSFH extracts in percent inhibition (%I) per 0.1 mg total protein will be determined. Data were not provided in percent inhibition, but rather in relative potency (RP), a value calculated using percent inhibition of pooled SC_D FSFH extract. The use of the pooled SC_D extract, which was comprised of equal volumes of all six SC_D FSFH extracts, as the control used to calculate RP confounded direct comparison of allergenic potency of GE vs. SC_D FSFH extracts when expressed as RP. The testing laboratory was unable to provide clarity regarding the calculation of RP values. (See discussion below for how CVM responded to this lack of clarity).

For valid comparisons among fish types, independence of evaluation and comparison should be maintained. ABT's use of combined SC_D FSFH extract to normalize all samples for assay and estimate RP compromised the ability to make independent comparisons among the fish types. A preferable strategy would be to use FSFH extract from farm raised fish.

ABT set its acceptability criteria for RP based on CBER's Allergen Vaccine Guidance. The release limits established by CBER for standardized dust mite and grass allergen vaccines of 0.5-2.0 are based on the performance characteristics of competition ELISAs for the determination of relative potency of these products. Because the RP values of 0.5-2.0 are release limits for standardized dust mite and grass allergen vaccine lots, we do not find these criteria relevant for our interpretation of this food safety study.

To allow an alternative evaluation, we requested all data and information relating to this study from ABT and the testing laboratory.

In our evaluation, we considered allergenic potency (in U/ml) at 40, 50, and 60% inhibition in lieu of RP values for two reasons. First, we wanted to compare directly the allergenic potency of GE diploid and triploid fish vs. sponsor control fish. This direct comparison was not possible using RP values that had been normalized using the pooled control FSFH, which was comprised of equal volumes of all six sponsor control FSFH extracts. Second, we were unable to determine how RP values for individual FSFH and pooled control FSFH extracts were calculated from allergenic potency at 40, 50 and 60% inhibition.

In each of the six assays for each FSHS, allergenic potency was estimated at 40, 50, and 60% inhibition. Although the differences between values provided an indication of measurement error and assay sensitivity, they did not provide information about variability that may exist between fish, which was one of our concerns. For initial evaluation, the mean allergenic potency for the six assay runs was estimated at 40, 50, and 60% inhibition. Additionally, using analysis of variance, we determined that the difference between the fish type (SC_D, TX_D and TX_T) was consistent whether measured at 40, 50, or 60 % inhibition. Because of this consistency, the final evaluation used the mean allergenic potency from all six assay runs estimated at all inhibition levels.

Initial evaluation of the results suggested that there may be an increase in the relative allergenic potency in the GE diploid salmon compared to sponsor control salmon. Given that salmon is often consumed as one individual fish fillet per serving rather than a mixture of many fish, we also considered the allergen level in individual fish in addition to group means.

Table 29 contains a summary of mean allergenic potency data.

Table 29. Summary of Mean Allergenic Potency of Salmon Extracts

UFID	Group	Ploidy	Mean
234	SC	Diploid	2.65
202	SC	Diploid	2.04
206	SC	Diploid	2.38
222	SC	Diploid	2.17
231	SC	Diploid	2.29
212	SC	Diploid	1.69
204	TX	Diploid	3.36
210	TX	Diploid	4.23
215	TX	Diploid	3.57
223	TX	Diploid	2.71
225	TX	Diploid	3.44
219	TX	Diploid	2.90
221	TX	Triploid	3.22
230	TX	Triploid	2.75
207	TX	Triploid	2.99
208	TX	Triploid	1.70
227	TX	Triploid	2.88
232	TX	Triploid	2.31

Although confidence in the data describing the diploid GE-salmon are low, these data indicated that four diploid GE fish had mean allergenic potency greater than 3.00 U/ml, with one fish having a mean allergenic potency value of 4.23 U/ml. Only one triploid GE salmon had a mean allergenic potency value greater than 3.00 U/ml. Figure 5 depicts the mean allergenic potency of individual fish.

Initial evaluation suggested that there may be an increase in the allergenic potency in the GE diploid salmon compared to sponsor control salmon. Allergenic potency (U/ml) data were analyzed using analysis of variance (ANOVA) with type included in the statistical model as a fixed effect. Pairwise comparisons of means for each of the TX groups were made to the mean for the SC_D group. P-values less than 0.05 indicate a statistically significant difference. There was no evidence of a statistically significant difference between the mean allergenic potency U/mL for sponsor control diploid (SC_D) fish compared to the triploid GE (TX_T) fish. A statistically significant difference existed between the mean allergenic potency U/mL for sponsor control diploid (SC_D) fish compared to the diploid GE (TX_D) fish. Table 30 summarizes the statistics for the mean allergenic potency per group.

Table 30. Least Squares Mean Allergenic Potency of Salmon Fillets per Group

Group*	Least Squares Mean Allergenic Potency	Standard Error	P-value from Test of Difference from Mean SC_D
SC_D	2.21	0.196	-
TX_D	3.37	0.196	0.0008
TX_T	2.64	0.196	0.1388

* SC_D = sponsor control non-GE diploid; TX_D = GE diploid; and TX_T = GE triploid.

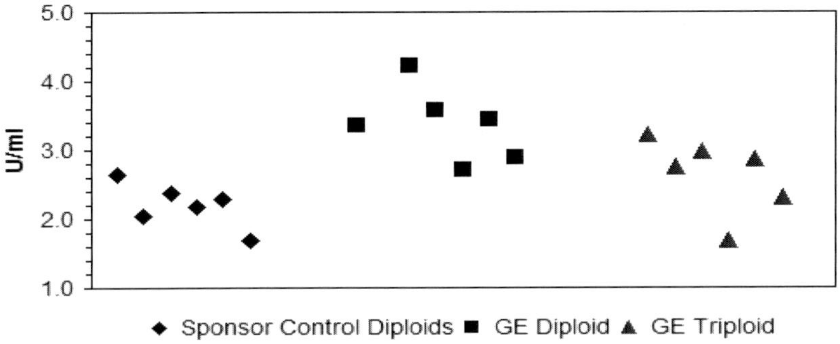

Black Diamonds: Sponsor Control Diploids; Black Squares: ABT diploid salmon; Black Triangles: ABT triploid salmon

Figure 5. Mean Allergenic Potency of Individual Fish.

Conclusion: The allergenic potency of triploid ABT salmon is not significantly different from that of sponsor control diploid salmon. There is insufficient data and information to draw a conclusion on the allergenic potency of diploid ABT salmon.

iv. Western Blot

The secondary objective of the study was to determine if any qualitative changes occurred in the major salmon allergen parvalbumin (*Sal s1*), due to the insertion of the AquAdvantage construct at the α- locus in ABT salmon. In this arm of the study, aqueous extracts from FSFH were analyzed using sodium dodecyl sulfate- polyacrylamide gel electrophoresis (SDS-PAGE) and Western blotting (immunoblotting).

The Western blot analyses were conducted in accordance with GLP. Validation information for assay methodology was provided. Acceptance specifications for the assay include the presence of a single band having an estimate between the 10 kD and 15 kD molecular weight marker for each lane containing an FSFH extract. Individual FSFH extracts were run twice, once each on two separate gels.

A mouse monoclonal anti-frog muscle parvalbumin antibody was used for Western blotting. Western blot membranes were documented by digital photography. Molecular weights of immunoreactive bands were calculated from migratory distances of the bands relative to a set of known molecular weight standards using image analysis software.

All Western blots exhibited only one single protein band of 11-13 kD reactive to anti-frog muscle parvalbumin antibody per lane. Based on these data, ABT concluded that both ABT salmon and non-GE Atlantic salmon express one predominant isoform of parvalbumin; therefore, there is no qualitative difference between parvalbumin expressed in ABT salmon and control Atlantic salmon.

Evaluation

We have determined that the technical flaws in this study so limit its interpretation that we can not rely on its results. These include a lack of appropriate controls, experimental conditions that preclude detection of more than one band per FSFH lane, and poor quality of the Western blots, and are described in greater detail below. That being said, we conclude that there are no biologically meaningful differences in the estimated molecular weights of the immunoreactive protein in GE salmon vs. sponsor control salmon detected by Western blot under assay conditions used. It is likely that the immunoreactive bands are parvalbumin, but our lack of confidence regarding the identity of bands limits our ability to draw a meaningful conclusion useful for public health purposes.

- Visual examination of digital photographs of Western Blots showed a single protein band of approximately 11-12 kD in each FSFH extract lane, consistent with the reported molecular weight of finfish parvalbumin. All bands were relatively weak; several bands were very faint. We note that for one run, the image of the Western blot appeared to be inverted with respect to the SDS- PAGE image and the figure legend. IBT was unable to confirm if the correct values were used to determine molecular weight.
- Regarding the use of an anti-frog parvalbumin antibody for Western blotting, frog and salmon parvalbumin are fairly evolutionarily conserved (Jenkins et al., 2007) and there is a high degree of immunological cross reactivity between frog and finfish parvalbumin (Hilger et al., 2004). In addition, the mouse monoclonal antibody to frog parvalbumin used in this study has been shown to detect finfish parvalbumin in immunoblots (Chen et al., 2006). Therefore, the selection of the anti-frog parvalbumin antibody was an acceptable choice for detection of salmon parvalbumin in the Western blots.
- Visual examination of the digital photographs of Sypro Orange-stained gels showed slight downward curve of dye front, or 'frowning', in the majority of gels. 'Frowning' is not an uncommon occurrence in PAGE, and it likely accounts for the minor differences in estimated molecular weight.
- Western blot analyses attempt to address only the molecular weight of parvalbumin, one major salmon allergen. No other salmon allergens were examined. Published literature has demonstrated a number of IgE-reactive bands in finfish extracts (putative finfish allergens) by Western blot using individual reactive human serum samples or pooled reactive human sera (Griesmeier et al., 2010; Hilger et al., 2004; Lim et al., 2008; Nakamura et al., 2009; Van Do et al., 2005). In addition, no attempt was made to quantitate the relative abundance of parvalbumin. No positive or negative controls were used; therefore immunoreactive bands cannot be identified as parvalbumin with a high degree of certainty. Finally, no comparisons can be made between different gels. Less abundant immunoreactive species could not be detected because of the acceptance criteria and experimental conditions that were established.

Conclusion: No reliable conclusions can be drawn from this study regarding parvalbumin in ABT salmon vs. non-GE control salmon.

v. Further Consideration of Allergenicity of GE Salmon

One potential indirect hazard that may result from the insertion of the AquAdvantage construct at the α- locus is a possible increase in the endogenous levels of allergens in ABT salmon due to insertional mutagenesis in a region of the genome that may act as a regulator of the expression of one or more of these proteins. Although the previous study attempted to address this point, its various technical deficiencies make it difficult to determine whether the allergenicity of salmon, or the prevalence of any known endogenous protein that has been implicated in allergic responses (i.e., parvalbumin) have changed, thereby somehow increasing the allergenicity of the fish.

Others attempted to address this issue. Nakamura *et al.* (2009) compared the allergenicity of growth hormone (GH) transgenic and non-transgenic amago salmon (*Oncorhynchus masou ishikawae*). Western blots using antibodies against frog parvalbumin and fish type-I collagen and 22 individual fish-allergic sera demonstrated no differences between GH-transgenic and non-transgenic amago salmon with respect to the amount of binding to known or suspected allergens. In this study, the rapid growth phenotype of a GH-transgenic salmon did not confer additional allergenicity as measured by *in vitro* IgE binding in this study.

Allergenicity conclusions

Triploid ABT salmon pose no additional allergenic risk than control Atlantic salmon. Insufficient data and information were available from which to draw a conclusion regarding possible additional allergenic risk posed by diploid ABT salmon.

c. Summary of and Conclusions from the Identification and Characterization of Indirect Food Consumption Hazards

Based on all previous criteria including statistical analyses, we conclude that levels of all analytes in ABT salmon are similar to levels in appropriate comparator salmon (e.g., either the sponsor controls, farmed salmon, literature reports, or some combination of the three).

We conclude that any differences observed for analytes are the result of normal biological variation and are highly unlikely to be associated with toxicological or nutritional hazards to humans consuming ABT salmon.

The statistically significant difference in mean vitamin B6 levels was investigated using a margin of exposure assessment. Even if the highest observed level of vitamin B6 observed in the diploid ABT salmon was to be found in all ABT salmon, the margin of exposure assessment indicated that it would still be well within the upper bound recommended daily intake for vitamin B6. We therefore find that there is no food consumption hazard due to vitamin B6.

For fatty acid analytes, values found in the ABT salmon were consistently more similar to the farmed control values than to the sponsor control values, and were proportional to total fat levels in these three groups of salmon; these differences did not appear to be due to differences in fat content or protein source of the ABT and sponsor control salmon diets.

ABT salmon offer a balanced diet of omega-3 and omega-6 fatty acids, similar to that provided by the FC controls and by farmed Atlantic salmon.

We conclude that ABT salmon are not materially different from other Atlantic salmon with respect to omega-3 and omega-6 fatty acid levels and ratio of omega-3 to omega-6 fatty acids.

Triploid ABT salmon pose no additional allergenic risk than control Atlantic salmon. Insufficient data and information were available from which to draw a conclusion regarding possible additional allergenic risk posed by diploid ABT salmon. No indirect food consumption hazards were identified.

D. Summary of the Characterization of Food Consumption Hazards

1. Identity

ABT salmon meet FDA's standard for identity for Atlantic salmon under the criteria established for the RFE. Therefore, we were able to conduct analyses to determine whether there were any changes in ABT salmon relative to non-GE Atlantic salmon, and if so, if any of those changes posed a food consumption hazard.

2. Summary of the Characterization of Direct Food Consumption Hazards

Direct food hazards have been defined as those resulting from exposure to the rDNA construct or its gene product. For the purposes of this evaluation, these include the Chinook salmon growth hormone, and other hormones in the somatotropic axis. We considered both results from the literature on GE salmon of the same parental lineage as the ABT salmon and ABT salmon themselves.

We looked for direct food consumption hazards. None were found.

- Sponsor's Study: Determination of IGF1, GH, T3, T4, 11-keto Testosterone, Testosterone, and Estradiol in Salmon Tissue.
  - No differences in the levels of growth hormone were observed in edible tissues.
  - Levels of estradiol, testosterone, 17-ketotestosterone, T3, and T4 were not different in the ABT salmon compared with comparator non-GE salmon.
- In order to determine whether an apparent difference in IGF1 in mature diploid ABT salmon posed a food consumption hazard, a margin of exposure (MOE) analysis was performed. That analysis, using conservative upper-bound assumptions regarding the presence of IGF1 in tissues, and conservative upper bound consumption rates found MOEs ranging from 508-1,220. We further noted that the value for the apparent elevation of IGF1 came from a diploid ABT salmon that is not the subject of this application for triploid female ABT salmon (AquAdvantage Salmon).
- The apparent difference in IGF1 in mature diploid ABT salmon compared to sponsor control non-GE salmon was relatively small. Analysis of levels of IGF1 via the MOE evaluation shows that the concentrations would not be biologically significant in the background of the existing systemic and dietary hormonal milieu for the consumer.
- The expression of Chinook salmon growth hormone in ABT salmon does not present a new risk of allergic reaction to salmon allergic individuals and is unlikely to cause allergic cross-reactions.
- Conclusions from Du et al. (1992).
  - Mean plasma growth hormone concentrations did not differ statistically between the GH transgenic Atlantic salmon (genetically engineered to contain the same construct as the ABT salmon) and either age-matched non-GE siblings or pooled control samples from age-matched siblings. Mean plasma T3 concentrations in the GH transgenic Atlantic salmon were statistically different from and lower in the GH transgenic Atlantic salmon compared to non-GE siblings but not when compared to the pooled controls. The highest plasma GH levels correlated with the largest GH transgenic Atlantic salmon while an inverse correlation was shown for mean plasma T3 concentrations.

3. Summary of the Characterization of Indirect Food Consumption Hazards

Indirect food consumption hazards have been defined as those arising from perturbations of the physiology of ABT salmon from the introduction of the rDNA construct or its gene product to alter the composition of food. In order to determine whether any indirect food hazards could be found, we evaluated the composition of ABT salmon, and performed an assessment of its allergenicity. No indirect food consumption hazards were found.

- The general composition (i.e., proximates, vitamins, minerals, and individual amino acids) ABT salmon did not differ in any biologically relevant way from its comparators when data from ABT-submitted studies were analyzed.
  - The statistically significant difference in mean vitamin B6 levels was investigated using a margin of exposure assessment. Even if the highest observed level of vitamin B6 observed in the diploid ABT salmon was to be found in all ABT salmon, the margin of exposure assessment indicates that it would still be well within the upper bound recommended daily intake for vitamin B6.
- No biologically relevant differences in the levels of individual fatty acids were observed between ABT salmon and its comparators.
- The amount of total fat in ABT salmon was similar to farmed control salmon, and within the 7-19% total fat lipid levels reported in the scientific literature for wild and farmed Atlantic salmon.
- The allergenic potency of triploid ABT salmon was not significantly different from that of sponsor control diploid salmon. Insufficient data and information exist to draw a conclusion on the allergenic potency of diploid ABT salmon.
- Triploid ABT salmon are not materially different from other Atlantic salmon based on their composition or allergenicity.

E. Characterization and Summary of Food Consumption Risks

We conducted a weight of evidence evaluation of the data and information presented in this application to assess the food safety of AquAdvantage Salmon. Primary deference was given to controlled studies submitted by ABT; data and information from the scientific literature were also considered for both the identification of potential hazards and for providing comparisons.

Because no food consumption hazards have been identified, there are no food consumption risks.

Historically, when the Office of New Animal Drug Evaluation has concluded that the use of a new animal drug is safe for human consumption, there was an assumption that animal byproducts derived from animals that use the drug are safe for animal feed. Therefore, a "food use" approval included "food and feed use." After evaluating the reviews for the AquAdvantage Salmon, including the molecular characterization of the GE animal lineage, the phenotypic characterization, genotypic and phenotypic durability, and the food and feed safety, the Division of Animal Feeds (DAF) has not identified any unique animal feed safety issues with respect to the introduction of the AquAdvantage construct into Atlantic salmon. We note that the evaluations and their corresponding data were generated from studies not specifically designed to examine animal feed safety. However, based on the evaluations of the Animal Biotechnology Interdisciplinary Group, and the subsequent determination that AquAdvantage Salmon is safe for human food use, DAF has not identified any safety issues for the use of AquAdvantage Salmon as animal feed.

F. Uncertainties

Although we have not identified any food consumption hazards for diploid AquAdvantage Salmon, because of the low quality of the study evaluating the allergenicity of salmon tissue, there are uncertainties regarding the allergenicity of edible products from diploid AquAdvantage Salmon.

G. Analytical Method for a Tolerance

A tolerance was not considered to be needed for residues resulting from insertion of the AquAdvantage construct into the AquAdvantage Salmon. Consequently, there was no need for the development of an analytical method for the tolerance.

H. Analytical Method of Identity

An analytical method for the purpose of identity of the GE animal containing the approved construct resulting from the insertion event and lineage evaluated for the NADA is reviewed in Food Safety Appendix A. It has been found to be acceptable.

I. Conclusions for Food Safety

ABT salmon meets the standard of identity for Atlantic salmon as established by FDA's Reference Fish Encyclopedia. All other assessments of composition have determined that there are no material differences in food from ABT salmon and other Atlantic salmon.

We conclude that food from the ***triploid*** ABT Salmon that is the subject of this application is as safe as food from conventional salmon, and that there is a reasonable certainty of no harm from consumption of food from triploid ABT salmon. No animal feed consumption concerns were identified.

Although we have found no food consumption hazards for ***diploid*** ABT salmon, due to uncertainties regarding the allergenicity of the tested tissue in a study of low quality, we recommend that if ABT wishes to introduce diploid ABT salmon into the food supply that they perform another study on the allergenicity of the diploid fish and submit it for evaluation as a supplement to this application.

VIII. Environmental Analysis

A. Overview

ABT has developed a line of GE Atlantic salmon with a rapid growth phenotype for use in commercial aquaculture (AquAdvantage Salmon). These fish contain an rDNA construct for expression of Chinook salmon growth hormone under the control of an ocean pout promoter. Although a number of lines of GE salmon were generated as part of the development process, ABT has limited its production to specific lines derived from the EO-1α founder containing the AquAdvantage construct at the α-locus; these lines are the sole focus of this analysis.

Working Product Definition

The working product definition of AquAdvantage Salmon employed in the EA and for the purposes of this evaluation is as follows:

> "A triploid Atlantic salmon (*Salmo salar*) bearing a single copy of the stably integrated a-form of the *opAFP-GHc2* gene construct at the α-locus in the EO-1α line, populations of which grow to a mean body weight of 100 g, and exhibit a significantly greater proportion of animals weighing 100 g or more, within 2700 deg-days of first-feeding when fed to satiety in water temperatures characteristic of present-day farming operations, that are produced as eyed-eggs for grow-out of triploid, hemizygous, all-female fish in physically-contained, production facilities."

Conditions of Production and Use

The product is intended for the land-based culture of Atlantic salmon for commercial sale and human consumption. For the current evaluation, the potential environmental risks of AquAdvantage Salmon were evaluated under the following specific conditions of production and use:

- Production of eyed eggs in Prince Edward Island (PEI), Canada;
- Shipment of eyed eggs to Panama;
- Grow-out of fish in the highlands of Panama;
- Processing of fish in Panama; and
- Shipment of table-ready processed fish to the United States (U.S.).

Any changes to the conditions of use and/or locations of production or grow-out (e.g., the addition of new grow-out facilities) will require further environmental evaluation, and if made after a decision to approve has been made, will likely trigger a requirement for filing a supplemental NADA.

Scope and Approach

These analyses have been conducted in accordance with principles and procedures described in Guidance 187 - Regulation of Genetically Engineered Animals Containing Heritable rDNA Constructs. The evaluation addresses only the environmental safety of AquAdvantage Salmon containing the AquAdvantage construct at the α locus. These analyses do not comment on the other data ABT provided unless directly relevant to environmental safety.

In addition, because the locations of egg production and grow out of AquAdvantage Salmon are outside of the United States, in accordance with Executive Order 12114, *Environmental Effects Abroad of Major Federal Actions*[15], in the EA and in this evaluation, consideration was given to environmental effects abroad including those to foreign countries not participating with the U.S. and not otherwise involved in the action, and to potential effects on the global commons outside the jurisdiction of any nation[16].

Prior to an approval decision being made for AquAdvantage Salmon, a Veterinary Medical Advisory Committee (VMAC) meeting will be held to solicit comments from appropriate outside experts (the VMAC members) and from the public. A public display copy of the EA will be released in conjunction with the VMAC meeting. A final decision to prepare a FONSI or an environmental impact statement (EIS) will be made after comments from the public and appropriate experts have been received and evaluated.

As specified in Section 1007 of the Food and Drug Administration Amendments Act of 2007 (FDAAA), the Commissioner of Food and Drugs shall consult with the Assistant Administrator of the National Marine Fisheries Service (NMFS) of the National Oceanic and Atmospheric Administration (NOAA) to produce a report on environmental risks associated with genetically engineered seafood products, including the impact on wild fish stocks. At this time, it is expected that the EA, once it has been found acceptable to support preparation of a FONSI, will form the basis for this report with respect to AquAdvantage Salmon. However, this could change if a decision is made to prepare an EIS rather than a FONSI. Preliminary discussions on this matter have already been held with NMFS/NOAA.

Given that risk mitigations in the form of several different types of containment or confinement (i.e., physical, biological, and geographical/geophysical), are or will be, in place at facilities used for the production and grow-out of AquAdvantage Salmon, this analysis focuses primarily on the adequacy and redundancy of these containment measures for their intended purposes to prevent escapes and reproduction (see Section VIII.D below). Information included in this evaluation comes from the EA, the PEI broodstock facility inspection, the site visit to the Panama grow-out facility, and the ABT study report on the induction of triploidy. Subsequently, the risk questions identified in Section VIII.B, above, are addressed to evaluate the potential for significant environmental effects to occur as a result of the NADA approval of AquAdvantage Salmon under the specified conditions of use (i.e., egg production in PEI, grow-out in Panama).

We have evaluated the EA document submitted by ABT and additional available information with respect to environmental safety including inspection and site visit records, and the method validation study for inducing triploidy. The information provided by ABT to evaluate the environmental safety of AquAdvantage Salmon is acceptable and complete. There is substantial, reliable information available in the environmental assessment document to conclude that GE Atlantic salmon in the EO-1α lineage that contain the AquAdvantage construct at the α-locus are not expected to have a significant impact on the quality of the human environment (1) in the United States; (2) in foreign nations not involved in the action; or (3) on the global commons when raised and reared under the current conditions of physical, biological, and geographical/geophysical confinement present at hatchery and grow-out facilities in Canada and Panama. Subject to further public and outside expert comment, there appears to be adequate justification at this time for preparation of a finding of no significant impact (FONSI). A final decision on whether to prepare a FONSI or an environmental impact statement (EIS) will be made after comments on the EA have been received from the public and appropriate outside experts and have been considered.

B. Risk Questions

Risk is the probability of harm, and harm, for the purposes of assessing the environmental risks of genetically engineered animals, "is defined as gene pool, species, or community perturbation resulting in negative impacts to community stability" (NRC, 2002). Further, risk is a product of the probability of exposure, $P(E)$, and the probability of harm given that exposure has occurred, $P(H|E)$, that is:

$$\text{Risk (R)} = P(E) \times P(H|E)$$

In assessing whether AquAdvantage Salmon present a risk to the environment, the evaluation addresses the chain of events that may result in exposure of the surrounding environment (as defined in the specific conditions of the NADA) to the GE salmon and potential harms (impacts) that may result from such exposures. See the general conceptual model for hazard characterization and risk assessment shown in Figure 6.

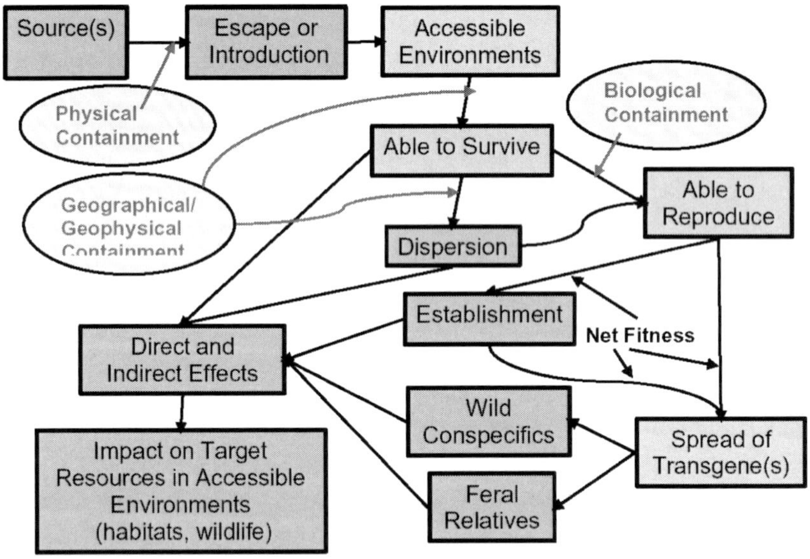

Figure 6. Conceptual Model for Hazard Characterization/Risk Assessment.

In general, exposure may be defined as establishment of a GE organism in the community (NRC, 2002). Thus for risk assessment, the critical factor is the likelihood the GE organism will become established in a community[17]. Considering this relationship and the general conceptual model for risk assessment shown in Figure 6, the following risk questions are addressed in this evaluation:

1. What is the likelihood that AquAdvantage Salmon will escape the conditions of confinement described in the EA and NADA?
2. What is the likelihood that AquAdvantage Salmon will survive and disperse if they escape the conditions of confinement described in the EA and NADA?
3. What is the likelihood that AquAdvantage Salmon will reproduce and establish if they escape the conditions of confinement as described in the EA and NADA?
4. What are the likely consequences to the surrounding environment, foreign nations not a party to the action, and the global commons should AquAdvantage Salmon escape the conditions of confinement as described in the EA and NADA?

C. Environmental Assessment Document

ABT submitted an EA addressing AquAdvantage Salmon dated January 11, 2010. This EA was prepared under the direction of Dr. Jane Staveley of the firm ARCADIS with assistance from ABT. The EA was certified by Dr. Ronald Stotish, President and CEO of ABT. A public display copy of the EA, which omits trade secret and confidential commercial information, has been prepared and will be made available for public review.

The following specific hazard-related questions were evaluated in the EA:

- Are GE salmon able to escape into the environment?
- If an accidental escape occurred, could GE salmon survive in the surrounding environment and compete with wild salmon (and escaped domestic non-GE salmon), or otherwise impact natural or ecological resources of global importance?
- Could the rDNA construct be transmitted to wild salmon, escaped non-GE domesticated salmon, or other species?
- Could GE salmon breed successfully with populations of wild salmon (and escaped domestic non-GE salmon)?
- Could the offspring resulting from these matings adversely affect the population of Atlantic salmon or other ecological resources of global importance?

Conclusions Stated in the Executive Summary of the EA

"The likelihood of escape, establishment, and spread of AquAdvantage Salmon is extremely small due to redundant containment measures, including physical, physicochemical, geographic/geophysical, and biological measures that are being implemented at the sites of egg production, grow-out and disposal. The combination of these various methods results in a very high degree of effective control. Physical measures include multiple mechanical means to prevent escape (e.g., screens, filters, etc.), while physico-chemical measures include the use of chlorine to kill any potential escapees. A strong management operations plan ensures that these containment measures are reliably implemented. Geographical and geophysical containment is provided by the location of the egg production and grow-out sites: the environment surrounding the egg-production site in Canada is inhospitable to early- life stages of Atlantic salmon due to high salinity; and, the environment downstream of the grow-out site in Panama is inhospitable to all life stages of Atlantic salmon due to high water temperatures, poor habitat, and physical barriers (e.g., several hydro-electric facilities). Biological containment is accomplished through the production of all-female triploid fish, which reduces the chance of breeding with native species, and significantly reduces the risk of transgene propagation in the environment."

"In summary, production and rearing of AquAdvantage Salmon will involve simultaneous, multiple, and redundant containment measures of various types that serve to mitigate the environmental risk quite adequately. These measures consist of producing triploid, all-female salmon that will be reared in land-based aquaculture systems possessing redundant physical containment measures engineered and managed to confine the fish to the culture systems and minimize the potential for escape. Furthermore, the facilities are located in geographical areas that are highly unfavorable to the survival, establishment and spread of AquAdvantage Salmon, should there be an escape."

"Consequently, the production, grow-out and disposal of AquAdvantage Salmon under the conditions described in this Environmental Assessment are highly unlikely to cause any significant effects on the environment."

Comments on the Adequacy of the EA

The EA is well written and effectively communicates information on the sites of egg production and grow-out of AquAdvantage Salmon, the phenotype of the fish, potential hazards (i.e., likelihood and consequences of escape, establishment and spread), and the containment measures that are being implemented to mitigate risks. Overall, the EA presents adequate documentation and analyses to support its conclusions. Comments and issues identified in reviews of earlier drafts of the EA were adequately addressed in the version prepared for public release[18].

Throughout the EA, triploid all-female AquAdvantage Salmon have been characterized as "sterile." This is potentially misleading because, as discussed at length in Section VIII.D.3.b of this document, sterility has not been explicitly verified in these fish and up to 5% of the eggs sold for grow-out may be non-triploid and still within release specifications. The effectiveness of triploidy in inducing sterility is discussed in Section 6.1.2.2 of the EA and is discussed further in Section VIII.D.3.b of this analysis. Based on research on other triploid fish, particularly females, we have reason to believe that the population of triploid, all-female AquAdvantage Salmon will be effectively sterile, with fertility greatly reduced or eliminated as a result of triploidy. However, it is recognized that a small proportion of the fish population, particularly those that are not triploid (i.e., ≤5%), may be fertile. In order to clarify the meaning of use of this term "sterile" in the EA, we requested ABT to explain by footnote or other means, the limitations on interpretation of this term at the first point in the EA where it is used in describing AquAdvantage Salmon. This has been done in the Introduction section of the public release version of the EA, which includes reference to Section 6.1 of the EA for further information on these issues.

Section 3.4 of the EA presents information on labeling of the product including several warning statements expected to appear on the product label: "fish must be reared in land-based, highly contained systems that prevent their release into the environment" and that the "fish cannot be reared in conventional cages or net pens deployed in open bodies of water." At the time that these analyses were prepared, labeling submitted for the product was still under review by FDA and thus is subject to change.

D. Evaluation of Containment and Environmental Effects

Conclusions in the EA that production and grow-out of AquAdvantage Salmon will not result in significant effects on the environment are based largely on the presence of a series of redundant containment measures applicable to both the salmon themselves and the facilities where they will be bred and reared. As explained in Section 8.2 of the EA, no single containment measure can be assured to be 100% effective. Therefore, optimum containment is dependent upon the deployment of several independent measures in series.

The U.S. Department of Agriculture's Agricultural Biotechnology Research Advisory Committee (USDA/ABRAC) has prepared performance standards for safely conducting research with genetically modified fish and shellfish (1995). These Performance Standards do not require, or even recommend, specific types and/or numbers of containment measures. With respect to risk management, the Performance Standards do state that although the number of independent containment measures[19] is site- and project-specific, they should generally range from three to five.

1. What is the likelihood that AquAdvantage Salmon will escape the conditions of confinement described in the EA and NADA?

The likelihood of escape depends primarily on the extent and adequacy of physical (mechanical) containment. Physical containment for egg production and grow-out is described in Sections 6.2.1 and 6.2.2 of the EA, respectively. Key components of physical containment for the PEI facility are indicated in Table 8 and Figure 8 of the EA. Similarly, for the Panama facility, key components are indicated in Table 9 and in Figure 9 of the EA. In addition, Section 6.2.3 describes the redundant, multi-level strategy used to insure containment at both facilities. Several Standard Operation Procedures (SOPs) are in place at the PEI facility to help insure containment, including an SOP addressing physical containment of GE salmonids. Figure 1 in this SOP, which is a schematic of the confinement equipment in place in the facility's early rearing area and grow-out area, has essentially been reproduced as Figure 8 of the EA.

a. Physical Containment at the PEI Facility

The adequacy of physical containment at the ABT PEI facility was addressed in the CVM evaluation of the 2001 EA prepared in support of investigational studies on AquAdvantage Salmon and in the facility site inspection conducted in October 2008 (see description below). All areas of the PEI facility have at least three independent forms of physical or mechanical containment. The areas of highest concern with respect to potential escape (i.e., egg incubation units and fry rearing tanks) have at least four to five separate, independent forms of physical containment.

Currently, eggs at the PEI facility are being incubated using Heath Stack incubators. When future production is scaled up, egg incubation is expected to occur in large (23 L) upwelling chambers instead of (or in addition to) the Heath Stack Incubators. Although not specifically addressed in the current EA, the physical containment conditions for these upwelling units will be equivalent to, or exceed, physical containment conditions currently in place for egg incubation.

b. FDA Inspection of PEI Broodstock and Hatchery Facility

An inspection of the ABT PEI broodstock and hatchery facility was conducted from October 7 - 9, 2008 by FDA as a limited directed inspection under CPGM 7368.001 (Preapproval inspections for NADAs). The FDA inspector was accompanied by three staff members of CVM's Office of New Animal Drug Evaluation. This was the initial pharmaceutical inspection of this facility. The facility was found to be in compliance with FDA regulations and no FDA483 form[20] was issued at the conclusion of the inspection.

Background:

An EA was submitted in December 2001 by ABT in support of the investigational use of AquAdvantage Salmon. This EA resulted in preparation of a FONSI by FDA for investigational studies under the INAD. Section 4.0 of the 2001 EA described the various passive and active forms of containment present at the ABT Canada facility in PEI, Canada. Passive containment includes physical-biological containment afforded by the surrounding environment (e.g., temperature, salinity, predators), while active containment describes the presence of physical barriers in the facility design (e.g., screens, nets) to prevent the escape or accidental release of fish and fish eggs to the outside environment.

Appendix IV of the 2001 EA contained SOPs in place at the facility relating to secure containment. The most relevant of the SOPs was addressing physical containment of GE salmonids. A key part of the SOP was Figure 1, a schematic of the confinement equipment in place in the facility's early rearing annex and grow out area, and the associated key to the components shown in this figure. The containment level (i.e., primary, secondary, etc.) for each component was described. According to the figure and key, all areas of the ABT facility have at least three independent forms of mechanical containment and some areas, including the egg incubation units and their discharges, have as many as four.

Actions and Findings:

During the site visit, the most recent copy version of SOP/ABPEI/2400 was requested. ABT provided a copy of version 2400.004, which was dated as effective on September 29, 2008. Figure 1 in this version of the SOP has been changed to reflect physical additions and modifications made to the facility several years ago, including enlargement of the early rearing area and changes in the sizes, shapes, and arrangement of tanks in certain parts of the facility. All areas of the facility were found to have at least two levels of containment and some have three or four[21]. Components shown and described in Figure 1 of the SOP that provide containment include the following:

Early Rearing Area

- Screened trays (egg incubators)
- PVC screening
- Catchment box & sock filters
- Containment sump with stainless steel perforated baskets (filters)
- Floor drain covers
- 60 micron drum filter and septic tank for solids removal
- Tank covers, slotted stand pipes, and overflow screens

Grow-Out Area

- External stand pipe screens
- Stand pipe covers
- Top nets or surround nets for each tank
- Floor drain covers (perforated steel plate; 1.5 or 7.0 mm)
- Chlorine puck in floor drain sump (during spawning of fish)
- Effluent containment sump with primary and secondary screening

The types and general locations of the containment components shown in Figure 1 of SOP 2400.004 were verified by visual inspection during a walk through of the PEI facility. Photographs were also taken of many of the key components. A detailed piping and instrument drawing (P&ID) was not available for the water/wastewater distribution system; therefore, it was not possible to verify the specific location and presence of each piece of equipment with a containment function. All components of the containment system that were observed appeared to be in good operational condition and functioning as designed.

Records ABT maintained relative to inspection of hatchery effluent screens and containment equipment indicated that these components were being inspected internally by ABT on a regular basis.

The Canadian governmental authorities charged with responsibility for the regulatory oversight of the research and development and the commercial deployment of transgenic aquatic organisms are Environment Canada and the Department of Fisheries and Oceans (DFO). Inspections of the facility by DFO occurred in 1996 and 2001. Reports from both DFO inspections found the facility "is as 'escape-proof' as one can reasonable expect." During the current inspection, a more recent DFO inspection report was requested. The FDA inspector was informed that the facility is no longer being inspected by DFO and that regulatory oversight in this area had essentially been turned over to Environment Canada.

c. Physical Containment at the Panama Grow-out Facility

The Panama grow-out facility includes small sizes of tanks for rearing fry and juveniles, plus large tanks for growing fish to market size (see Figure 9 in EA). The fry tanks contain either interior or exterior stand pipes, plus a series of two to three mechanical fine mesh screens (1 – 1.5 mm for small fry; 3 – 12 mm for larger fry and juveniles) made of metal to prevent fish from escaping. In addition, all water from these tanks must pass through a 500 micron sock filter prior to entering a drainage canal that collects all water from the facility and sends it to a series of four settling ponds (and from there to a nearby river). Thus, at a minimum, three levels of physical containment are present for these early life stages of AquAdvantage Salmon.

Grow-out (production) tanks have external stand pipes (to control the water height) and drain water through a slotted (0.9 cm), rigid PVC drainage plate in the tank bottom. The drainage plate and slots serve as the primary form of physical containment for the fish in these tanks.

From the grow-out tanks, water is routed to the drainage canal that also collects water from the fry tanks and other facility operations. There are two additional mechanical (6 and 12 mm) screens within a concrete containment sump that filter water from the drainage canal prior to it entering the series of four settling ponds. There is also a 12 mm rigid metal screen on the outlet of each of the four ponds. These larger screens would act as effective barriers to larger fry, juveniles and adults, but are not expected to preclude passage of small fry (or eggs). Taken as a whole, and counting the series of settling ponds with screens as only a single form, there are four independent forms of physical containment applicable to fish reared in the grow-out tanks.

Although not present at the time of the CVM site visit in November 2009, egg incubation units will be put in place after the NADA approval of AquAdvantage Salmon occurs and commercial-scale production is initiated. According to ABT, physical containment conditions for the incubation units is expected to be similar to those currently in place for egg incubation at the PEI facility, offering a minimum of four levels of containment.

Additional containment in the way of tank netting and chain link security fences is present to limit access by potential predators and unauthorized personnel.

Information reported in the EA with respect to the Panama facility was verified during the site visit conducted by CVM staff in mid-November, 2009 (see below). Multiple forms of physical (mechanical) containment were present and as described in the EA. In addition, the facility appeared to be newly built and well-maintained.

d. Site Visit of the ABT Grow-out Facility in Panama

From November 10-12, 2009, a site visit of the ABT grow-out facility in Panama was conducted by two CVM staff members along with a fisheries scientist from NOAA/NMFS. This site visit was conducted primarily to verify that the conditions of rearing and containment at the grow-out facility are as described in the EA, and to evaluate any other factors which would influence the potential for escape. A secondary objective of the visit was to observe and gain information on the local environment, including portions of the river adjacent and downstream of the grow-out facility, to help ascertain whether AquAdvantage Salmon would be likely to survive and establish should they somehow, in fact, escape the grow-out facility.

Based on observations made and information gathered during the site visit, the descriptions and schematics provided in the EA on the Panama grow-out facility, the river and surrounding environment have been accurately represented. There are a minimum of three or four levels of containment between both the fry tanks and grow-out tanks and the river. This includes counting the series of four downstream settling ponds (each with its own outlet screen) as only one level of containment.

Visual observations of the river adjacent to the ABT grow-out facility indicate a very high gradient profile with high current velocity and substrate consisting predominately of large rocks and boulders. Except in terms of water temperature, the river habitat in the vicinity of the ABT facility does not appear to be favorable to Atlantic salmon, or most other fish species for that matter, although it would not necessarily preclude survival and possibly establishment (if salmon were reproductively competent). Populations of rainbow trout are reported to occur in the river as a result of intentional stocking by the Panamanian government as far back as 1925 (Section 4.2.2 of EA), however, the abundance of these trout has not been well documented and they were not observed by CVM staff during the site visit.

e. Issues Affecting Containment and Security

Natural Disasters

The USDA/ABRAC Performance Standards state that the siting and location of physical facilities must prevent accidental releases of GE fish and shellfish during floods, storms, earthquakes, and other natural disasters. When possible, facilities should be located above the 100-year flood level, and storm drains should be designed to a 100-year rainfall event, or storage provided. If this is not possible, measures should be taken to insure that surface runoff is diverted around the facility, or that other measures are in place to effectively minimize the potential for release of animals.

The potential for accidental releases of AquAdvantage Salmon at the production and grow-out sites due to natural disasters has not been explicitly addressed in the EA except that it is stated that "no such natural disasters have occurred, or are known to occur, in proximity to the PEI and Panama facilities." In addition, issues relating to facility location have not been discussed in relation to the USDA/ABRAC Performance Standards. The production site on PEI is potentially subject to hurricanes and strong winter storms and may fall within the 100-year flood level. The facility is located approximately 25 feet above sea level at its highest point and approximately 120 feet from a tidal river. Although close to this river and the Gulf of St. Lawrence, there are barriers across much of the river mouth at its confluence with the Gulf, which is approximately one mile away, thus it is highly unlikely that storm surges would directly impact the facility or subject it to flooding.

The grow-out facility in Panama is potentially subject to flooding conditions from a nearby river. The area receives a significant amount of annual rainfall, approximately 570 cm or 224 inches per year (Table 4 in EA), with much of it coming in the wet summer months. There was a significant flood of the river in the recent past that caused extensive damage at locations downstream of the grow-out facility. The facility itself, however, was not directly affected by flood waters and there was no serious damage. The only incidental damage was sustained as a result of debris that clogged the metal intake screens filtering water from the river as it enters the concrete water distribution canal. In the time since this accident occurred, redundant intake piping has been added and many of the pipes have been moved underground to prevent future occurrences of this type. Considering that this flooding was among the worst to ever occur in the area, it seems improbable that the grow-out facility would be impacted by future events of this type in a manner that could cause accidental release of GE fish.

Physical Security

The USDA/ABRAC Performance Standards call for security measures to (a) control normal movement of authorized personnel, (b) prevent unauthorized access to the site, and (c) for outdoors projects, eliminate access of predators that could potentially carry animals offsite. The Performance Standards also mention the possible need for alarms, stand-by power, and an operational plan (including training, traffic control, record keeping, and an emergency response plan). The current EA does not address most of these issues in any detail. For example, Tables 8, 9 and 10 of the EA list information with respect to physical security measures at the Panama grow-out facility, but there is little or no additional discussion in the text of these measures. Additional information was requested of ABT in this regard and further details were provided in a subsequent correspondence and added to the public display copy of the EA. Based on the information provided by ABT, as well as observations made by

CVM personnel during the PEI facility inspection and Panama facility site visit, the physical security measures in place appear to be adequate to address the concerns listed in the USDA/ABRAC Performance Standards. ABT is aware that unauthorized access to these sites may represent a potential hazard and has taken appropriate steps to reduce the possibility this will occur.

Transportation of Eggs from PEI to Panama

The potential for the unintentional release of AquAdvantage eggs during transport from the Canada to the grow-out facility has not been explicitly discussed in the EA. Section 3.4 of the EA briefly describes shipping from Canada to Panama as occurring via air freight with subsequent ground-shipment to the grow-out facility by ABT personnel. ABT has been requested to include a brief discussion of the potential for escape/release under this transportation scenario in a subsequent version of the EA, including the final public display version. Because of the reproductive containment measures in place for AquAdvantage Salmon (triploidy and all-female populations), any escape or unintentional release is not expected to result in adverse consequences.

f. Disposal of Fish and Fish Wastes

Disposal of AquAdvantage Salmon (mortalities and processing wastes) and waste products from these fish is discussed in Section 3.3 of the EA for both the PEI and Panama facilities. In PEI, mortalities and culls requiring disposal will be stored frozen until they are incinerated. In Panama, fish mortalities will be deposited in 1-m deep, on- site burial pits. Individual fish will be separated by a layer of caustic lime and the pit will be filled with soil once it has reached a depth of 0.5 m with mortalities. Fish wastes from the PEI facility are subject to extensive treatment prior to discharge to the local estuary. In Panama, wastes from the grow-out tanks will be removed from the facility's effluent in a series of four sedimentation ponds prior to discharge to a nearby river.

Fish processing (i.e., production of fillets) will occur at a processing plant that is located within a short drive of the grow-out facility. AquAdvantage Salmon will be sacrificed at the grow-out facility, placed on ice, and then transported to the processing plant for filleting. The method by which the fish wastes generated through processing (i.e., heads, bones, and entrails) will be disposed was not described in the EA, except that disposal will be "in accordance with applicable laws." FDA has no regulatory jurisdiction over disposal of wastes in other countries. This is only a concern in that it represents a potential exposure route for rDNA constructs.

No specific hazards or risks have been identified in conjunction with mortalities and fish wastes. The integrated EO-1α construct is not inherently hazardous and is not expected to be mobilized through waste disposal. The only potential hazard associated with processing wastes is the possibility for incorporation into animal feed. Any specific risks associated with this exposure route will be evaluated as part of the food/feed safety evaluation.

Conclusion: The probability that AquAdvantage Salmon will escape from either the PEI egg production facility or the Panama grow-out facility is extremely small due to the presence of multiple, independent forms of physical (mechanical) containment at both facilities. This containment has been verified by FDA through an inspection (PEI) and site visit (Panama).

Physical security and containment is acceptable at both sites to insure that it is unlikely there will be any unintentional releases of salmon due to natural disasters or malicious activities.

2. What is the likelihood that AquAdvantage Salmon will survive and disperse if they escape the conditions of confinement described in the EA and NADA?

The likelihood of survival and dispersal of AquAdvantage Salmon, given that escape occurs, depends primarily on their phenotype (e.g., tolerance to physico-chemical parameters such as temperature and dissolved oxygen) and aspects of geographical and geophysical containment present as a result of the specific location and surrounding environment. We define geographical and geophysical containment as the presence of inhospitable conditions in the surrounding environment that would preclude or significantly reduce the probability of survival, dispersal, and/or long-term establishment should an animal escape confinement at its site of rearing. Several aspects of this type of containment apply, at least in part, to AquAdvantage Salmon at both the production and grow-out sites. Unless deemed to be 100% effective under all reasonably foreseeable circumstances, containment of this type would normally be considered to be secondary to other types. The applicability of this type of containment to AquAdvantage Salmon is discussed in Section 6.3 of ABT's EA and separately below for the PEI broodstock and Panama grow-out facilities.

a. Geographical/Geophysical Containment for the PEI Facility

The PEI facility lies on the southern shore of a tidal river close to its confluence with the Gulf of Lawrence (Atlantic Ocean). Water from the facility, including effluent from all floor drains, fish tanks and egg incubators, eventually discharges to this river. Environmental conditions in the vicinity of the facility are generally conducive to adult Atlantic salmon, but not to early life stages of these fish (eggs, fry and pre-smolts), particularly at the time of year fish would be spawned at the facility - November and December. Water temperatures in the winter months are typically very low (less than 0 °C) and the water has a relatively high salinity, in the range of 21 parts per thousand (ppt)[22]. Therefore, it is highly unlikely that early life stages of any Atlantic salmon at the facility would be able to survive if they were able to escape the multiple levels of physical containment in place. The same is less likely to be true for older fish; however, it is still unlikely they would be able to survive the sudden transition from a low salinity, freshwater environment, to a moderately high salinity, brackish water environment.

b. Geographical/Geophysical Containment for the Panama Facility

The Panama facility lies at an elevation of approximately 5,000 feet with water supplied by nearby spring. The temperature of the spring water is fairly constant throughout the year and at approximately 15 °C, is similar to that of the river that runs next to the facility and receives its water discharges. This temperature is near the optimum for Atlantic salmon growth and would not be an impediment to survival should any eggs or fish escape from the facility. In fact, populations of rainbow trout, a related fish species which also requires fairly low water temperatures and high dissolved oxygen concentrations, are reported to occur in the river in general vicinity of the ABT grow-out facility.

However, as indicated by temperature data in Table 3 of the EA, the temperature of the nearby river increases substantially as it drops in elevation, merges with another river downstream, and the combined flow approaches the Pacific Ocean. In the lower reaches of the watershed, the water temperature is in the range of 26 to 28 °C, at or near the upper incipient lethal level[23] for Atlantic salmon, which is approximately 28 °C for acclimated juveniles according to a study conducted by Elliott (1991) (see discussion in Section 2.4.1.4 of the EA). In Elliot's study, feeding stopped when the water temperature exceeded 22.5 °C; therefore, it is expected that long-term survival would be compromised due to starvation at locations even further upstream of those which would be more directly lethal due to temperature alone. As a result, it is extremely unlikely that AquAdvantage Salmon would ever be able to survive and migrate to the Pacific Ocean. In addition, because surface water temperatures in the Pacific Ocean along the Panamanian coast are in the range of 25 to 28°C throughout the year (National Oceanic Data Center, online data for 2009)[24], survival of these salmon in the ocean is also extremely unlikely.

We are unaware of studies on the temperature tolerance of AquAdvantage Salmon. There is no reason, however, to believe the upper tolerance (i.e., upper incipient lethal limit) would be higher for AquAdvantage Salmon than for non-GE Atlantic salmon. GH- transgenic salmon have been reported to have an increased requirement for dissolved oxygen (D.O.) compared to non-GE counterparts (see Section 2.4.3.2 of EA), which is presumably related to their faster rate of growth. This could indicate a reduced tolerance to higher water temperatures as the D.O. content of water at saturation is inversely related to water temperature. Based on the studies of Stevens et al. (1998) on GH-transgenic salmon, which showed that D.O. in the water starts to become limiting as the level dropped to 6 mg/L (ppm), oxygen alone would not appear to be limiting for AquAdvantage Salmon if they were to reach the lower reaches of the watershed. At a temperature of 26 to 28°C, the D.O. content of the river water would be at approximately 8 mg/L if at or near a saturation level well above the limiting concentration of 6 mg/L. The lowest levels of D.O. levels in the river basin are 7.0 to 7.2 mg/L based on water quality monitoring over the years 2002-2008 (see Table 6 of EA). This suggests that D.O. would likely not be a factor limiting survival.

A significant amount of the water volume of the river adjacent to ABT's Panama grow-out facility may be diverted for use in local hydroelectric power plants. The hydroelectric power plants and associated water diversion dams appear to constitute significant, although not complete, barriers to fish movement within the watershed, particularly with respect to potential downstream migration of AquAdvantage Salmon to the Pacific Ocean.

In addition to high water temperatures, several other conditions of the aquatic habitat in the lower sections of the watershed are also not favorable for salmonids. For example, the amount of solids in the water column is high (Table 6 in EA) and the macroinvertebrate fauna, while diverse, are not abundant, thus food availability may be limited. There is also the potential for predation by a resident population of introduced rainbow trout; the distribution of these trout in the watershed is unknown.

c. Fitness of AquAdvantage Salmon

In our phenotypic characterization of AquAdvantage Salmon (see Section V above), several phenotypic changes were identified in published scientific articles on growth hormone (GH) transgenic Atlantic salmon[25]. The phenotypic changes described are consistent with the presence of the EO-1α construct and appear to result in decreased fitness (e.g., increased oxygen requirements, decreased critical swimming speed, lower metabolic scope, etc.). The observed changes in phenotype were expected to reduce the chances for survival and establishment should any AquAdvantage Salmon escape from commercial production facilities.

Conclusion: The geographical and geophysical conditions present in the aquatic environments near both locations will act to limit the potential survival and spread of AquAdvantage Salmon to other locations. This is particularly true for the eggs and early life stages of these salmon in PEI, which are unlikely to survive if exposed to high salinity and low temperature conditions in nearby aquatic environment, and for all life stages of these salmon in Panama, which are unlikely to survive the high temperature conditions in the lower reaches of the watershed.

3. What is the likelihood that AquAdvantage Salmon will reproduce and establish if they escape the conditions of confinement described in the EA and NADA?

The likelihood that AquAdvantage Salmon will reproduce and establish depends largely on the extent and adequacy biological containment (bioconfinement). Information ABT submitted to us and contained in the EA with respect to bioconfinement is summarized and discussed below.

Biological Containment (Bioconfinement)

a. Validation of Triploidy Method

ABT submitted a study conducted to determine the validity of the method and conditions used for the production of triploid Atlantic salmon by ABT at its Aquaculture Research and Development Facility on Prince Edward Island, Canada. The primary objective of the study was to determine if the conditions for induction of triploidy using hydrostatic pressure treatment could be employed in a reproducible manner for the batch-wise production of triploid eggs during the commercial manufacture of AquAdvantage Salmon.

During the study, one-to-one crosses were established with eggs from non-GE female Atlantic salmon fertilized with milt from AquAdvantage salmon males that were hemizygous for EO-1α. The fertilized eggs from each cross were apportioned volumetrically into five replicate groups: one diploid control group that was not pressure treated, and four treated replicates that were subjected to hydrostatic pressure shock (9500 psi for five minutes at 300 deg-min post-fertilization).

After treated, fertilized eggs had developed to the 'eyed' stage (~325-400 deg day), 350 eyed-eggs were arbitrarily sub-sampled to estimate the proportion of triploid individuals in the aggregate population. Ploidy analysis was performed on sub-samples of homogenates of a pool of 10 eyed-eggs collected from each of the four treated replicates from five different independent crosses (i.e., a total of 20 independent pressure shocked groups). Determination of ploidy was determined using a flow cytometer with samples from the diploid control groups serving as a reference standard. Results of the method validation study are summarized in Table 10 (Section VI, Genotypic and Phenotypic Durability).

Based on the analysis of ploidy in all 20 replicates, the average proportion of triploids produced from the five independent crosses was 99.8%. For individual treatment events (i.e., all replicates), the proportion of triploidy ranged from 98.9% to 100%. Triploidization was very similar for each of the five independent crosses, on average ranging from only 99.7% to 99.9%. The lowest effectiveness observed for an individual batch of eggs was 98.9%.

b. Triploidy, Triploidization, and Sterilization of AquAdvantage Salmon

AquAdvantage Salmon eggs distributed for grow-out will be subjected to pressure treatment shortly after fertilization to induce triploidy. Information on triploidy, the reliability of the method used by ABT to induce triploidy, and the effectiveness of triploidy in inducing sterility is contained in Section 6.1 of the EA and has been summarized above. As reported in the EA (Section 6.1.2.1), laboratory-scale efficiencies of 100% have been reported for inducing triploidy in Atlantic salmon in the literature, but less than 100% effectiveness is expected on a commercial scale with larger batch sizes. Data generated in the ABT validation study indicate an effectiveness of 99.8% on average.

As part of the Durability Plan to which ABT has committed, testing will continue to be conducted on all composite batches of fertilized eggs that are intended to be sold commercially. If triploidization in these eggs is not greater than 95% (based on the 95% lower confidence limit), the entire batch of eggs must be destroyed. (We note again that the lowest effectiveness observed for an individual batch of eggs was 98.9%.) Because the testing methodology results in egg destruction, it would be impossible to ensure 100% triploidy in eggs sold for grow-out.

AquAdvantage Salmon are described throughout the EA as being "sterile." We assume that this assertion is based on the common characterization in the scientific literature that "triploid" equals "sterile." Although adequate demonstration of triploidy has been provided, ABT has not submitted any specific data to show whether or not AquAdvantage Salmon are indeed sterile. Although perhaps not completely sterile, the fertility of triploid females is expected to be greatly reduced by the procedure in light of information discussed below.

Information in Section 6.1.2.2 of the EA (*Effectiveness of triploidy in inducing sterility*) has been presented on the subject and the article on triploid fish by Benfey (1999) also addresses the issue. These sources suggest that it is highly likely that triploid Atlantic salmon, particularly female salmon, will be effectively sterile due to failure of the gametes to mature normally. Most germ cells do not progress through the first meiotic prophase in triploids of either sex; however, Benfey (1999) cites several studies that have reported the occasional production of mature oocytes by triploid females. According to Benfey, it is likely that triploid females are able to produce small numbers of mature, post-meiotic cells, but that their growth occurs at such a slow rate that they are not observed at the normal time of sexual maturation in diploids. Benfey suggests this slow growth would help explain the great variability in egg size observed in triploid salmonids, suggestive of asynchronous development.

There are reports in the literature of coho salmon and other unidentified fish species that are triploid and sometimes fertile (Lee and Donaldson, 2001). In addition, as cited in the EA, Gillet et al. (2001) have reported on research in Arctic char which indicates that some triploid females are able to develop ovaries and mature gametes. However, fecundity was low in these females (approximately one third of that in diploids) and their fertilized eggs did not hatch.

Another potential issue is that of mosaic individuals. A small percentage of putative triploids can become mosaic – bearing both diploid and triploid cells – as has been found in studies with fish and oysters (NRC, 2004). Bioconfinement would be compromised if cells within gonadal tissue were mosaic. Although it cannot be completely ruled out, no published data showing this occurrence in fish were found in a search by the NRC (NRC, 2004). Considering the available data, we conclude that the triploidy will greatly reduce, if not totally eliminate, fertility in ABT's female AquAdvantage Salmon.

c. Female, Mono-Sex Populations

As described and illustrated in Section 3.1.1 of the EA, ABT is using a complex production process involving gynogenesis and neomales (sex-reversed females)[26] to insure that a monosex, all female population of AquAdvantage Salmon is produced for grow-out. Producing monosex populations is another effective form of reproductive containment to complement that of triploidy, but there are other potential advantages in producing all-female populations rather than those of males only. Triploid males are still capable of exhibiting spawning behavior with fertile females, which could potentially lead to decreased reproductive success for these females[27]. In addition, the degree of gonadal development in triploid females is believed to be less than that in triploid males (Arai, 2001; Benfey, 1999; Thorgaard and Allen, 1992), presumably due to the absence of vitellogenic oocyte growth, which in turn is believed to be due to a lack of ovarian estrogen synthesis (Benfey, 1999). Many studies have reported the production of spermatozoa by triploid male fish, but at greatly reduced numbers, resulting in very dilute milt (Benfey, 1999). In addition, the spermatozoa from triploids are generally aneuploid and when the milt from triploid males is used to fertilize normal haploid eggs, the resultant progeny typically begin development but die at embryonic and larval stages (Benfey, 1999).

The effectiveness of the methods used by ABT to insure that an all-female population of AquAdvantage Salmon is produced has not specifically been evaluated quantitatively in any studies to date.

> Using gynogenesis[28] as part of the process, rather than chemically-induced sex-reversal alone, not only eliminates the time and labor that would be needed to distinguish neomales from true males following androgen treatment, but also essentially insures 100% effectiveness in producing a genetically all-female population with a full complement of maternal DNA. Because gynogenesis has only been used relatively recently by ABT as part of their production process, and because of the importance of this process in insuring an all-female population of AquAdvantage Salmon, ABT has made a commitment to conducting additional genotypic post-approval monitoring of the AquAdvantage Salmon neomales as part of its Durability Plan.

Conclusion: A minimum of 95% of the AquAdvantage Salmon eggs sold for commercial production use will be triploid and 100% are expected to be female. The expected average percentage of triploidy is 99+% based on the method validation study. The fertility of triploid females is expected to be greatly reduced or negligible compared to diploid females. The combination of triploidy and an all-female population is expected to render AquAdvantage Salmon effectively sterile and to result in complete reproductive containment due to the absence of males.

4. What are the likely consequences to the surrounding environment, foreign nations not a party to the action, and the global commons should AquAdvantage Salmon escape the conditions of confinement described in the EA and NADA?

The possibility for the action (i.e., FDA approval of AquAdvantage Salmon with egg production in Canada and grow-out in Panama) to result in effects on the global commons, foreign nations not a part to the action, and stocks of threatened and endangered species is addressed in Sections 7.1, 7.2, and 7.3, of the EA, respectively. No significant effects or risks were identified taking into consideration the containment and confinement measures currently in place for the fish and facilities. In addition, the proposed containment conditions for AquAdvantage Salmon are consistent with guidelines from the North Atlantic Salmon Conservation Organization (NASCO) in its "Williamsburg Resolution." This resolution calls for rearing of transgenic (i.e., GE) salmon in secure, self-contained, land-based facilities.

Section 8 of the EA presents additional information with respect to risk assessment for AquAdvantage Salmon. Based on concerns expressed by members of the Ecological Society of America (Snow et al., 2008), Table 11 in the EA summarizes the likelihood of potential risks for six major environmental processes and their potential ecological consequences. In the table, it is concluded that AquAdvantage Salmon present no significant risk to any of these processes.

Conclusion: There is adequate information in the EA to address the potential consequences of escape of AquAdvantage Salmon on the surrounding environment, foreign nations not participating in the action, and global commons (including stocks of wild Atlantic salmon).

None of this information suggests that escape of AquAdvantage Salmon would result in significant effects on the environment with egg production in PEI and grow-out in the Panamanian highlands.

E. Addressing the Risk Questions

Risk Question 1: What is the likelihood that AquAdvantage Salmon will escape the conditions of confinement described in the EA and NADA?

There are multiple and redundant physical and mechanical barriers in place in the water systems at the PEI egg production and Panama grow-out facilities to prevent the accidental release of eggs and/or fish to nearby aquatic environments. These barriers have been designed specifically to prevent the escape of different life stages of AquAdvantage Salmon. Both facilities have a minimum of three to five mechanical barriers in place for all internal flow streams which release water to the environment. This level of containment is consistent with recommendations in the ABRAC Performance Standards and has been verified by an FDA inspection or site visit. Therefore, the likelihood is considered very low that AquAdvantage Salmon will escape from confinement at these sites. In addition, physical security and containment is acceptable at both sites to insure that it is unlikely there will be any unintentional releases of salmon due to natural disasters or malicious activities.

Risk Question 2: What is the likelihood that AquAdvantage Salmon will survive and disperse if they escape the conditions of confinement described in the EA and NADA?

Should they escape, the likelihood of survival of AquAdvantage Salmon is dependent on the life stage(s) escaping and the location. Information on the phenotype of these fish suggests their fitness may be reduced compared to non-GE Atlantic salmon; however, fitness is not expected to be compromised to such an extent that survival is greatly affected, at least on a short-term basis. An exception would be eggs and early life stages (i.e., all pre-smolt stages) at the PEI facility. They would not be expected to survive the conditions of salinity (and temperature depending on the time of year) in the accessible environments of PEI if they escape confinement. It is also possible that older life stages (juveniles to adult) of AquAdvantage Salmon will not be able to survive if they escape confinement in PEI and enter nearby estuarine and marine environments.

There are no specific study data addressing this issue for AquAdvantage Salmon, but information in the scientific literature indicates that when hatchery-reared salmon smolts are retained in fresh water, they will "desmolt" and lose their tolerance to salinity (Lundqvist et al., 1986; McCormick et al., 1998). If true for AquAdvantage Salmon, this loss of salinity tolerance would be expected to result in rapid death if these salmon were to enter the local tidal river (estuary) or nearby ocean.

Freshwater-reared GE coho and Chinook salmon containing growth hormone rDNA constructs apparently do not lose their ability to tolerate high levels of salinity, even as adults (Robert Devlin, Canadian Department of Fisheries and Oceans, personal communication, August 13, 2010). If this is also true for AquAdvantage Salmon because this tolerance is a secondary result of the added growth hormone gene, survival of adults and older life stages would not be expected to be compromised if escape occurred in PEI. In the absence of specific data for juvenile and adult AquAdvantage Salmon on survivability under high salinity conditions, we have conservatively assumed that older life stages of these salmon would survive if they escape containment in PEI.

Should escape of AquAdvantage Salmon occur in Panama, survival is only expected in the vicinity of the grow-out facility and upper watershed of the adjacent river. High temperature conditions and water diversion projects further downstream will limit the long-term survival of all life stages of AquAdvantage Salmon; therefore, long-range dispersal is precluded. Survival outside of freshwater conditions in Panama (i.e., in the Pacific Ocean) is considered impossible due to high water temperature conditions.

Risk Question 3: What is the likelihood that AquAdvantage Salmon will reproduce and establish if they escape the conditions of confinement as described in the EA and NADA?

Because of the multiple forms of biological (reproductive) containment inherent to AquAdvantage Salmon to be sold in commerce, namely triploidy and all-female populations, these fish are effectively sterile and cannot reproduce. This essentially precludes establishment of a population of these fish in the accessible environments should escape occur. The only potential means for establishment (or pseudo-establishment) are through the escape of reproductively competent broodstock at the PEI facility or through a continual series of escapes at the Panama facility. Neither of these scenarios is likely given the physical containment measures in place at both facilities and both would require the escape of a significant number of animals, something even less likely. Therefore, given the available information, it is concluded that the likelihood is extremely small that AquAdvantage Salmon will establish and reproduce if they escape from facilities in PEI and Panama.

Risk Question 4: What are the likely consequences to the surrounding environment, foreign nations not a party to the action, and the global commons should AquAdvantage Salmon escape the conditions of confinement as described in the EA and NADA?

Should AquAdvantage Salmon escape physical confinement in PEI or Panama, there are no likely consequences on the U.S., foreign nations not participating in the action, or on the global commons as a result of applicable reproductive and geographic/geophysical confinement. There is only one realistic scenario that potentially could lead to adverse outcomes outside of the local environment. That would be the escape (or intentional malicious release) of a large number of reproductively competent broodstock from the PEI egg production facility. These fish could potentially survive in the local environment depending on the time of year escape were to occur; however, although native to PEI, there are no longer significant runs of natural Atlantic salmon on the island. On PEI, the main limitation to salmon production is believed to be due to stream sedimentation and barriers to migration (see section 4.1.2 of the EA). Future returns of Atlantic salmon in PEI rivers are expected to remain dependent on stocking of hatchery-reared fish. Given these circumstances, it is highly unlikely that escaped fish from ABT's PEI facility would be able to establish in the local environment (or farther afield) and cause any significant impacts. The number of broodstock in the PEI facility is quite limited; therefore, the potential for the mass release of thousands of fish, as sometimes occurs during net-pen farming of Atlantic salmon, is impossible. Therefore, it is concluded that there are no likely consequences or effects expected on the U.S., any foreign nations not participating in the action, or the global commons. In addition, no effects on stocks of threatened or endangered species, including stocks of wild Atlantic salmon, are expected.

The possibility for effects to occur on the local environment in Panama, and to a much lesser extent in PEI, cannot be totally ruled out if AquAdvantage Salmon were to escape. Except to the extent that they might eventually cause impacts on the U.S., foreign nations not a party to the action, and the global commons, these potential effects do not need to be evaluated under NEPA and Executive Order 12114. The ability of Canada and Panama to evaluate these potential effects and/or regulate AquAdvantage Salmon is not preempted by any FDA actions and remains under local jurisdiction.

F. Conclusion

We have evaluated the EA document submitted by ABT and additional available information with respect to environmental safety including inspection and site visit records, and the method validation study for inducing triploidy. From all indications, adequate containment measures appear to be in place at both the PEI and Panama facilities to insure a very low probability of escape for all life stages of salmon present. At both facilities, there are multiple and redundant physical and mechanical barriers in place in the water systems to prevent movement of eggs and/or fish to nearby aquatic environments; a minimum of three to five mechanical barriers for all flow systems in both facilities, which is in line with recommendations in the ABRAC Performance Standards. In addition, the geographic and geophysical conditions in the aquatic environments present near both locations would further limit potential survival and spread of AquAdvantage Salmon to other locations. This is particularly true for eggs and early life stages of these salmon in PEI, which are unlikely to survive if exposed to high salinity and low temperature conditions in the nearby estuary or the Gulf of St. Lawrence, and for all life stages of these salmon in Panama, which are unlikely to survive the high temperature conditions in the lower reaches of the watershed. Further, because of redundancy in reproductive containment (i.e., triploidy and all-female populations), any escapees in Panama that are able to survive in the local conditions would not be able to reproduce and permanently establish themselves in the local environment. As a result of all of these containment measures, the potential occurrence of any significant effects on the global commons or any foreign nations not participating in this action is considered extremely remote. In addition, no effects on stocks of wild Atlantic salmon are expected.

Except for minor issues to be addressed in the final public display version of the EA, the information provided by ABT to evaluate the environmental safety of AquAdvantage Salmon is acceptable and complete. There is substantial, reliable information available in the environmental assessment document to conclude that GE Atlantic salmon in the AquAdvantage lineage that contain the AquAdvantage construct at the α-locus are not expected to have a significant impact on the quality of the human environment (1) in the United States; (2) in foreign nations not involved in the action; or (3) on the global commons when raised and reared under the current conditions of physical, biological, and geographic/geophysical confinement present at hatchery and grow-out facilities in Canada and Panama. Subject to further public and outside expert comment, there appears to be adequate justification at this time for preparation of a finding of no significant impact (FONSI).

A final decision on whether to prepare a FONSI or an environmental impact statement (EIS) will not be made until after comments on the EA have been received from the public and appropriate outside experts and have been considered.

IX. Claim Validation

The Claim Validation step focuses on whether the GE animal meets those characteristics claimed in the Product Definition. The description and the statements provided in the Product Definition serve as the basis of the Claim Validation evaluation and drive the collection of the data. The studies and the extent of the data required for the Claim Validation evaluation are unique to each application. In addition, the Claim Validation evaluation may draw on the data and conclusions from other steps of the hierarchical review process.

In the Product Definition of this application, ABT claims that AquAdvantage Salmon grow to a mean body weight of at least 100 g, and exhibit a significantly greater proportion of animals weighing 100 g or more, within 2700 degree-days of first-feeding when fed to satiety in water temperatures characteristic of present-day farming conditions. Generally speaking, this means that AquAdvantage Salmon reach 100 g, which is a size that is important to those in the aquaculture industry, sooner than conventionally farmed Atlantic salmon.

The evaluation of this claim was conducted in accordance with principles and procedures described in Guidance 187 - Regulation of Genetically Engineered Animals Containing Heritable rDNA Constructs.

Data provided by ABT demonstrate that when compared to diploid unmodified salmon, (1) the AquAdvantage Salmon grow to a mean body weight of at least 100 g within 2700 degree-days of first-feeding, and (2) a greater proportion of AquAdvantage Salmon grow to at least 100 g within 2700 degree-days after first-feeding under normal commercial aquaculture conditions.

The data presented in support of this claim are based on the fish from the 2007 year class. The summary of this data is presented in the Figure 7 below.

Statistical review of the data showed that the presence of the AquAdvantage construct in fish had a significant effect on the fish body weight and proportion of fish weighing more than 100 g at 2700 degree-days. A summary of results is presented in Table 31 below.

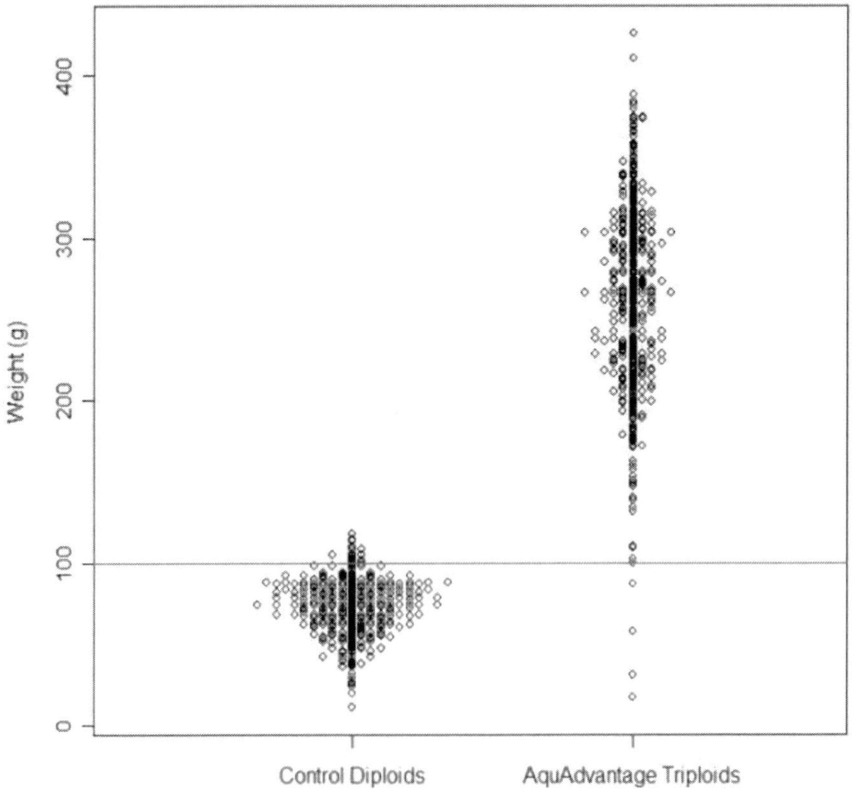

Figure 7. Weight of AquAdvantage Salmon and Comparators at 2700 deg C days.

Table 31. Comparisons of Comparator Diploid Salmon with AquAdvantage Salmon

	Number Of Fish	Weight (g)		Fish Weighing > 100 g	
		Mean	Standard Error	Number	Percent
Control Diploids	306	72.6	1.02	15	4.9
AquAdvantage Triploids	369	261.0	3.29	364	98.6

Triploid AquAdvantage Salmon reached a mean weight significantly different from the mean weight reached by the diploid control group (261.0 g versus 72.6 g, respectively; $p < 0.0001$). Additionally, the percentage of triploid AquAdvantage Salmon exceeding 100 g at 2700 degree-days (98.6%) is significantly different from that of diploid controls (98.6% versus 4.9 %, respectively; $p < 0.0001$). Because the triploid AquAdvantage Salmon show significant results compared to the diploid controls, we can conclude that the AquAdvantage construct is responsible for the observed phenotype.

REFERENCES

Abrahams & Sutterlin. (1999). The foraging and antipredator behaviour of growth-enhanced transgenic Atlantic salmon. *Animal Behaviour*, *58*, 933-942.

Arai, K. (2001). Genetic improvement of aquaculture finfish species by chromosome manipulation techniques in Japan. *Aquaculture*, *197*, 205-228.

Astwood, J. D., Leach, J. N. & Fuchs, R. L. (1996). Stability of food allergens to digestion in vitro. *Nature Biotechnology*, *14*, 1269-1273.

Benfey, T. J. (1999). The physiology and behavior of triploid fishes. *Reviews in Fisheries Science*, *7*, 39-67.

Benfey, T. J. (2001). Use of sterile triploid Atlantic salmon (Salmo salar L.) for aquaculture in New Brunswick, Canada. *ICES Journal of Marine Science*, *58*, 525.

Cal, R. M., Vidal, S., Camacho, T., Piferrer, F. & Guitian, F. J. (2005). Effect of triploidy on turbot haematology. *Comparative Biochemistry and Physiology-Part A: Molecular & Integrative Physiology*, *141*, 35-41.

Center for Food Safety and Applied Nutrition (CFSAN), Food and Drug Administration. (2006). Revised report: Approaches to Establish Thresholds for Major Food Allergens and for Gluten in Food. Available at http://www.fda.gov/Food/LabelingNutrition/FoodAllergensLabeling/GuidanceComplianceRegul atoryInformation/ucm106108.htm.

Chafen, J. J. S., Newberry, S. J., Riedl, M. A., Bravata, D. M., Maglione, M., Suttorp, M. J., Sundaram, V., Paige, N. M., Towfigh, A. & Hulley, B. J. (2010). Diagnosing and managing common food allergies: a systematic review. *JAMA*, *303*, 1848.

Chen, L., Hefle, S. L., Taylor, S. L., Swoboda, I. & Goodman, R. E. (2006). Detecting fish parvalbumin with commercial mouse monoclonal anti-frog parvalbumin IgG. *J Agric Food Chem*, *54*, 5577-5582.

Codex Alimentarius Commission (CAC). (2008). Guideline for the Conduct of Food Safety Assessment of Foods Derived from Recombinant-DNA Animals (Codex Alimentarius Commission). CAC/GL 68-2008. Available at http://www.codexalimentarius.net/download/standards/11023/CXG_068e.pdf.

Cogswell, A. T., Benfey, T. J. & Sutterlin, A. M. (2001). The hematology of diploid and triploid transgenic Atlantic salmon (Salmo salar). *Fish Physiology and Biochemistry*, *24*, 271-277.

Cook, J. T., McNiven, M. A., Richardson, G. F. & Sutterlin, A. M. (2000a). Growth rate, body composition and feed digestibility/conversion of growth-enhanced transgenic Atlantic salmon (Salmo salar). *Aquaculture*, *188*, 15-32.

Cook, J. T., McNiven, M. A. & Sutterlin, A. M. (2000b). Metabolic rate of pre-smolt growth-enhanced transgenic Atlantic salmon (Salmo salar). *Aquaculture*, *188*, 33-45.

Deitch, E. J., Fletcher, G. L., Petersen, L. H., Costa, I., Shears, M. A., Driedzic, W. R. & Gamperl, A. K. (2006). Cardiorespiratory modifications, and limitations, in post-smolt growth hormone transgenic Atlantic salmon Salmo salar. *Journal of Experimental Biology*, *209*, 1310.

Dorafshan, S., Kalbassi, M. R., Pourkazemi, M., Amiri, B. M. & Karimi, S. S. (2008). Effects of triploidy on the Caspian salmon Salmo trutta caspius haematology. *Fish Physiology and Biochemistry*, *34*, 195-200.

Du, S. J., Gong, Z., Fletcher, G. L., Shears, M. A., King, M. J., Idler, D. R. & Hew, C. L. (1992a). Growth enhancement in transgenic Atlantic salmon by the use of an "all fish" chimeric growth hormone gene construct. *Nature Biotechnology*, *10*, 176-181.

Du, S. J., Gong, Z., Hew, C. L., Tan, C. H. & Fletcher, G. L. (1992b). Development of an all-fish gene cassette for gene transfer in aquaculture. *Molecular marine biology and biotechnology*, *1*, 290.

Elliott, J. M. (1991). Tolerance and resistance to thermal stress in juvenile Atlantic salmon, Salmo salar. *Freshwater Biology*, *25*, 61-70.

Evans, M. L. & Neff, B. D. (2009). Non-additive genetic effects contribute to larval spinal deformity in two populations of Chinook salmon (Oncorhynchus tshawytscha). *Aquaculture*, *296*, 169-173.

Food Allergen Labeling and Consumer Protection Act of 2004 FALCPA. (2004). Food Allergen Labeling and Consumer Protection Act of 2004. P.L. 108-282.

Food and Drug Administration (FDA). (1992). Statement of Policy: Foods Derived from New Plant Varieties. 57 Federal Register 22984. Available at http://www.fda.gov/Food/GuidanceComplianceRegulatoryInformation/GuidanceDocuments/Biotechnology/ucm096095.htm.

Food and Drug Administration (FDA). (1993). Freedom of Information Summary. POSILAC (sterile sometribove zinc suspension). For increasing production of marketable milk in lactating dairy cows. Original Approval.

Fjelldal, P. G., Glover, K. A., Skaala, Imsland, A. & Hansen, T. J. (2009). Vertebral body mineralization and deformities in cultured Atlantic salmon (Salmo salar L.): Effects of genetics and off-season smolt production. *Aquaculture*, *296*, 36-44.

Fletcher, G. L., Hew, C. L. & Davies, P. L. (2001). Antifreeze proteins of teleost fishes. *Annual Review of Physiology*, *63*, 359.

Friesen, E. N., Ikonomou, M. G., Higgs, D. A., Ang, K. P. & Dubetz, C. (2008). Use of terrestrial based lipids in aquaculture feeds and the effects on flesh organohalogen and fatty acid concentrations in farmed Atlantic salmon. *Environ Sci Technol*, *42*, 3519-3523.

Frost, R. A. & Lang, C. H. (2003). Regulation of insulin-like growth factor-I in skeletal muscle and muscle cells. *Minerva endocrinologica*, *28*, 53.

Galbreath, P. F. & Thorgaard, G. H. (1995). Saltwater performance of all-female triploid Atlantic salmon. *Aquaculture*, *138*, 77-85.

Gendel, S. M. (1998). Sequence databases for assessing the potential allergenicity of proteins used in transgenic foods. *Advances in food and nutrition research*, *42*, 63-92.

Gillet, C., Vauchez, C. & Haffray, P. (2001). Triploidy induced by pressure shock in Arctic charr (Salvelinus alpinus): growth, survival and maturation until the third year. *Aquating Living Resources*, *14*, 327-334.

Gong, Z. & Hew, C. L. (1993). Promoter analysis of fish antifreeze protein genes. *Biochemistry and Molecular Biology of Fishes*, *2*, 307–324.

Goodman, R. E., Vieths, S., Sampson, H. A., Hill, D., Ebisawa, M., Taylor, S. L. & Van Ree, R. (2008). Allergenicity assessment of genetically modified crops—what makes sense? *Nature Biotechnology*, *26*, 73-81.

Griesmeier, U., Vázquez-Cortés, S., Bublin, M., Radauer, C., Ma, Y., Briza, P., Fernández-Rivas, M. & Breiteneder, H. (2010). Expression levels of parvalbumins determine allergenicity of fish species. *Allergy*, *65*, 191-198.

Haffray, P., Aubin, J., Houis, V., Labbe, L. & Jalabert, B. (2007). Comparison of pressure or thermal treatments on triploid yields and malformations up to swim up stage in rainbow trout (Oncorhynchus mykiss). *Aquaculture*, *272*, S265.

Hamilton, M. C., Hites, R. A., Schwager, S. J., Foran, J. A., Knuth, B. A. & Carpenter, D. O. (2005). Lipid composition and contaminants in farmed and wild salmon. *Environ Sci Technol*, *39*, 8622-8629.

Hammer, R. E., Pursel, V. G., Rexroad, C. E., Wall, R. J., Bolt, D. J., Ebert, K. M., Palmiter, R. D. & Brinster, R. L. (1985). Production of transgenic rabbits, sheep and pigs by microinjection.

Hefle, S.L., Nordlee, J.A. & Taylor, S.L. (1996). Allergenic foods. *Critical Reviews in Food Science and Nutrition*, *36*, S69-89.

Hew, C. L., Wang, N. C., Joshi, S., Fletcher, G. L., Scott, G. K., Hayes, P. H., Buettner, B. & Davies, P. L. (1988). Multiple genes provide the basis for antifreeze protein diversity and dosage in the ocean pout, Macrozoarces americanus. *Journal of Biological Chemistry*, *263*, 12049.

Hilger, C., Thill, L., Grigioni, F., Lehners, C., Falagiani, P., Ferrara, A., Romano, C., Stevens, W. & Hentges, F. (2004). IgE antibodies of fish allergic patients cross-react with frog parvalbumin. *Allergy*, *59*, 653-660.

Hobbs, R. S. & Fletcher, G. L. (2008). Tissue specific expression of antifreeze protein and growth hormone transgenes driven by the ocean pout (Macrozoarces americanus) antifreeze protein OP5a gene promoter in Atlantic salmon (Salmo salar). *Transgenic research*, *17*, 33-45.

Ikonomou, M. G., Higgs, D. A., Gibbs, M., Oakes, J., Skura, B., McKinley, S., Balfry, S. K., Jones, S., Withler, R. & Dubetz, C. (2007). Flesh quality of market-size farmed and wild British Columbia salmon. *Environ Sci Technol*, *41*, 437-443.

Institute of Medicine of the National Academies (IOM). (2001). Dietary Reference Intakes (Washington, DC, Institute of Medicine (IOM) of the National Academies, Food and Nutrition Board).

Jaenisch, R. & Mintz, B. (1974). Simian virus 40 DNA sequences in DNA of healthy adult mice derived from preimplantation blastocysts injected with viral DNA. Proceedings of the National Academy of Sciences of the United States of America, *71*, 1250.

Jenkins, J. A., Breiteneder, H. & Mills, E. N. (2007). Evolutionary distance from human homologs reflects allergenicity of animal food proteins. *Journal of Allergy and Clinical Immunology, 120*, 1399-1405.

Johnson, R. M., Shrimpton, J. M., Heath, J. W. & Heath, D. D. (2004). Family, induction methodology and interaction effects on the performance of diploid and triploid Chinook salmon (Oncorhynchus tshawytscha). *Aquaculture, 234*, 123-142.

Juskevich, J. C. & Guyer, C. G. (1990). Bovine growth hormone: human food safety evaluation. *Science, 249*, 875.

Kennedy, S. R., Campbell, P. J., Porter, A. & Tocher, D. R. (2005). Influence of dietary conjugated linoleic acid (CLA) on lipid and fatty acid composition in liver and flesh of Atlantic salmon (Salmo salar). *Comparative Biochemistry and Physiology Part B: Biochemistry and Molecular Biology, 141*, 168-178.

Knapp, G., Roheim, C. A. & Anderson, J. L. (2007). The great salmon run: competition between wild and farmed salmon. TRAFFIC North America.

Kvellestad, A., Hoie, S., Thorud, K., Trud, B. & Lyngoy, A. (2000). Platyspondyly and shortness of vertebral column in Atlantic salmon Salmo salar in Norway–description and interpretation of pathologic changes. *Diseases of Aquatic Organisms, 14*, 97–108.

Lee, C. S. & Donaldson, E. M. (2001). General discussion on "Reproductive biotechnology in finfish aquaculture". *Aquaculture, 197*, 303-320.

Levesque, H. M., Shears, M. A., Fletcher, G. L. & Moon, T. W. (2008). Myogenesis and muscle metabolism in juvenile Atlantic salmon (Salmo salar) made transgenic for growth hormone. *Journal of Experimental Biology, 211*, 128.

Lim, D. L. C., Neo, K. H., Yi, F. C., Chua, K. Y., Goh, D. L. M., Shek, L. P. C., Giam, Y. C., Van Bever, H. P. S. & Lee, B. W. (2008). Parvalbumin–the major tropical fish allergen. *Pediatric Allergy and Immunology, 19*, 399-407.

Lundqvist, H., Clarke, W. C., Eriksson, L. O., Funegård, P. & Engstrøm, B. (1986). Seawater adaptability in three different river stocks of Baltic salmon (Salmo salar L.) during smolting. *Aquaculture, 52*, 219-229.

McCormick, S. D., Hansen, L. P., Quinn, T. P. & Saunders, R. L. (1998). Movement, migration, and smolting of Atlantic salmon (Salmo salar). *Canadian Journal of Fisheries and Aquatic Sciences, 55*, 77-92.

Nakamura, R., Satoh, R., Nakajima, Y., Kawasaki, N., Yamaguchi, T., Sawada, J., Nagoya, H. & Teshima, R. (2009). Comparative study of GH-transgenic and non-transgenic amago salmon (Oncorhynchus masou ishikawae) allergenicity and proteomic analysis of amago salmon allergens. *Regulatory Toxicology and Pharmacology, 55*, 300-308.

National Institute of Allergy and Infectious Diseases (NIAID). (2010). Draft Guidelines for the Diagnosis and Management of Food Allergy, N.I.o.A.a.I. Diseases, ed. Available at http://www.niaid.nih.gov/topics/foodAllergy/clinical/Documents/guidelines.pdf.

Nordlee, J. A., Taylor, S. L., Townsend, J. A., Thomas, L. A. & Bush, R. K. (1996). Identification of a Brazil-nut allergen in transgenic soybeans. *New England Journal of Medicine. 334*, 688.

NRC. (2002). *Animal biotechnology: science-based concerns*, D. Davis, ed.

O'Flynn, F. M., McGeachy, S. A., Friars, G. W., Benfey, T. J. & Bailey, J. K. (1997). Comparisons of cultured triploid and diploid Atlantic salmon (Salmo salar L.). *ICES Journal of Marine Science*, *54*, 1160.

Ørnsrud, R., Gil, L. & Waagbø, R. (2004). Teratogenicity of elevated egg incubation temperature and egg vitamin A status in Atlantic salmon, *Salmo salar L. Journal of Fish Diseases*, *27*, 213-223.

Palmiter, R. D., Brinster, R. L., Hammer, R. E., Trumbauer, M. E., Rosenfeld, M. G., Birnberg, N. C. & Evans, R. M. (1982a). Dramatic growth of mice that develop from eggs microinjected with metallothionein–growth hormone fusion genes.

Palmiter, Y., Richard, D. & Brinster, R. L. (1982b). Differential regulation of metallothionein- thymidine kinase fusion genes in transgenic mice and their offspring. *Cell*, *29*, 701-710.

Piferrer, F., Beaumont, A., Falguière, J. C., Flajshans, M., Haffray, P. & Colombo, L. (2009). Polyploid fish and shellfish: Production, biology and applications to aquaculture for performance improvement and genetic containment. *Aquaculture*, *293*, 125-156.

Polvi, S. M. & Ackman, R. G. (1992). Atlantic salmon (Salmo salar) muscle lipids and their response to alternative dietary fatty acid sources. *Journal of Agricultural and Food Chemistry*, *40*, 1001-1007.

Roberts, R. J., Hardy, R. W. & Sugiura, S. H. (2001). Screamer disease in Atlantic salmon, Salmo salar L., in Chile. *Journal of Fish Diseases*, *24*, 543-549.

Sadler, J., Pankhurst, P. M. & King, H. R. (2001). High prevalence of skeletal deformity and reduced gill surface area in triploid Atlantic salmon (Salmo salar L.). *Aquaculture*, *198*, 369-386.

Sampson, H. A. (2004). Update on food allergy. *Journal of Allergy and Clinical Immunology*, *113*, 805-819.

Saunders, R. L., Fletcher, G. L. & Hew, C. L. (1998). Smolt development in growth hormone transgenic Atlantic salmon1. *Aquaculture*, *168*, 177-193.

Shears, M. A., Fletcher, G. L., Hew, C. L., Gauthier, S. & Davies, P. L. (1991). Transfer, expression, and stable inheritance of antifreeze protein genes in Atlantic salmon (Salmo salar). *Molecular marine biology and biotechnology*, *1*, 58-63.

Snow, A. A. Andow, D. A., Gepts, P., Hallerman, E. M., Power, A., Tiedje, J. M. & Wolfenbarger, L. L. (2008). *Genetically Engineered Organisms and the Environment: Current Status and Recommendations1.*

Stevens, E. D. & Sutterlin, A. (1999). Gill morphometry in growth hormone transgenic Atlantic salmon. *Environmental Biology of Fishes*, *54*, 405-411.

Stevens, E. D., Sutterlin, A. & Cook, T. (1998). Respiratory metabolism and swimming performance in growth hormone transgenic Atlantic salmon. *Canadian Journal of Fisheries and Aquatic Sciences*, *55*, 2028-2035.

Stevens, E. D., Wagner, G. N. & Sutterlin, A. (1999). Gut morphology in growth hormone transgenic Atlantic salmon. Journal of Fish Biology *55*, 517-526.

Tamashiro, K. L. K., Wakayama, T., Akutsu, H., Yamazaki, Y., Lachey, J. L., Wortman, M. D., Seeley, R. J., D'Alessio, D. A., Woods, S. C. & Yanagimachi, R. (2002). Cloned mice have an obese phenotype not transmitted to their offspring. *Nature Medicine*, *8*, 262-267.

Thorgaard, G. H. & Allen, S. K. (1992). Environmental impacts of inbred, hybrid, and polyploid aquatic species. *Dispersal of living organisms into aquatic ecosystems*, 281.

Torstensen, B. E., Bell, J. G., Rosenlund, G., Henderson, R. J., Graff, I. E., Tocher, D. R., Lie, O. & Sargents, J. R. (2005). Tailoring of Atlantic salmon (Salmo salar L.) flesh lipid composition and sensory quality by replacing fish oil with a vegetable oil blend. *J Agric Food Chem*, *53*, 10166-10178.

United States Department of Agriculture (USDA). (1995). Performance Standards for Safely Conducting Research with Genetically Modified Fish and Shellfish, U.S.D.A., ed. (Agricultural Biotechnology Research Advisory Committee), pp. 156.

United States Department of Commerce (USDOC) Technology Administration NTIS. (2002). Foods Commonly Eaten in the United States: Quantities Consumed per Eating Occasion and In A Day, 1994-1996.

Ungemach, F. R. (1998). Recombinant Bovine Somatotropins (addendum). Toxicological Evaluation of Certain Veterinary Drug Residues in Food (IPCS WHO).

Vågsholm, D. (1998). Risk factors for spinal deformities in Atlantic salmon, *Salmo salar L. Journal of Fish Diseases*, *21*, 47-53.

Van Do, T., Elsayed, S., Florvaag, E., Hordvik, I. & Endresen, C. (2005). *Allergy* to fish parvalbumins: studies on the cross-reactivity of allergens from 9 commonly consumed fish. *Journal of allergy and clinical immunology*, *116*, 1314-1320.

Walker, S. C., Christenson, R. K., Ruiz, R. P., Reeves, D. E., Pratt, S. L., Arenivas, F., Williams, N. E., Bruner, B. L. & Polejaeva, I. A. (2007). Comparison of meat composition from offspring of cloned and conventionally produced boars. *Theriogenology*, *67*, 178-184.

Watanabe, S. & Nagai, T. (2008). Health status and productive performance of somatic cell cloned cattle and their offspring produced in Japan. *Journal of Reproduction and Development*, *54*, 6-17.

Watanabe, S. & Nagai, T. (2009). Death losses due to stillbirth, neonatal death and diseases in cloned cattle derived from somatic cell nuclear transfer and their progeny: a result of nationwide survey in Japan. *Animal Science Journal*, *80*, 233-238.

WHO-FAO. (2004). Vitamin and Mineral Requirements in Human Nutrition (Geneva, Switzerland, World Health Organization (WHO) and the Food and Agriculture Organization of the United Nations (FAO), 175-179.

Witten, P. E., Gil-Martens, L., Huysseune, A., Takle, H. & Hjelde, K. (2009). Towards a classification and an understanding of developmental relationships of vertebral body malformations in Atlantic salmon (Salmo salar L.). *Aquaculture 295*, 6-14.

Yaskowiak, E. S., Shears, M. A., Agarwal-Mawal, A. & Fletcher, G.L. (2006). Characterization and multi-generational stability of the growth hormone transgene (EO-1alpha) responsible for enhanced growth rates in Atlantic Salmon. *Transgenic research*, *15*, 465-480.

XI. Appendices

A. Phenotypic Characterization Appendix[29]

Hematology Values[30]

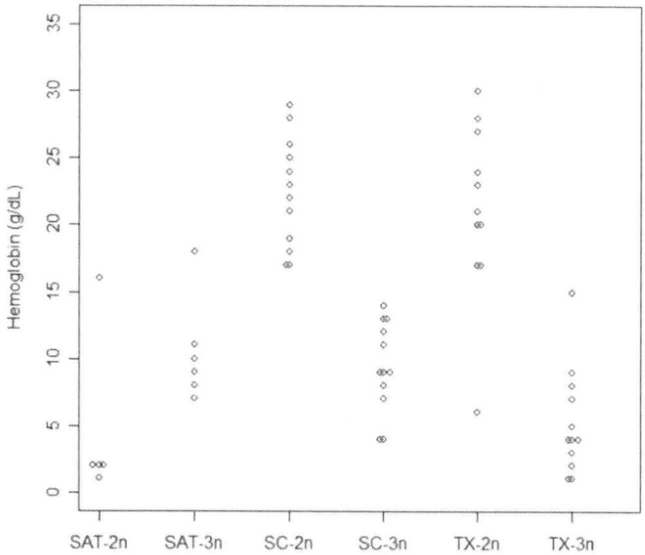

Figure 1. Hemoglobin.

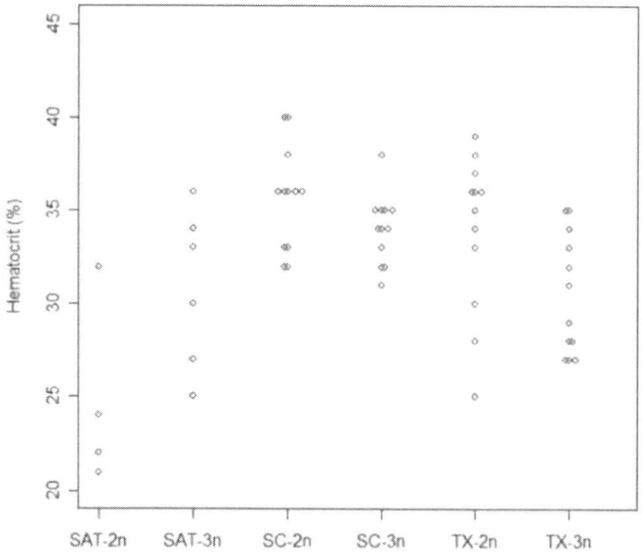

Figure 2. Hematocrit.

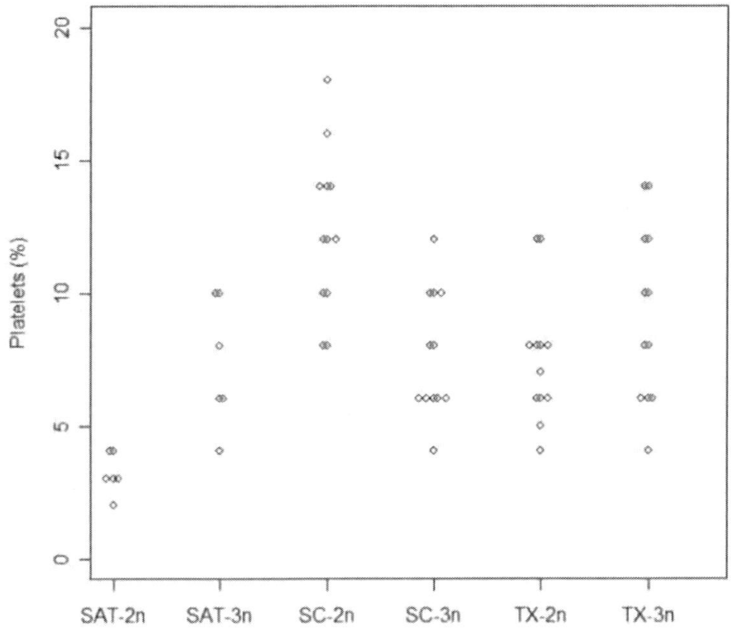

Figure 3. Platelets.

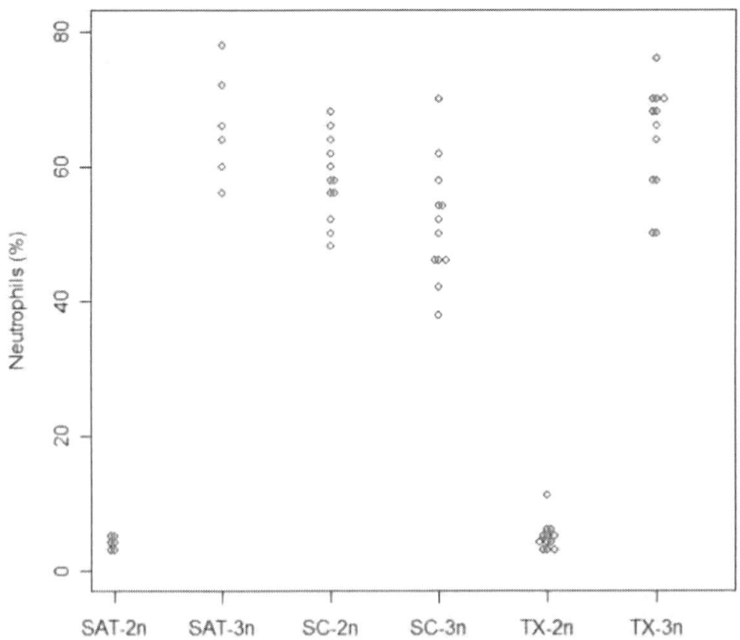

Figure 4. Neutrophils.

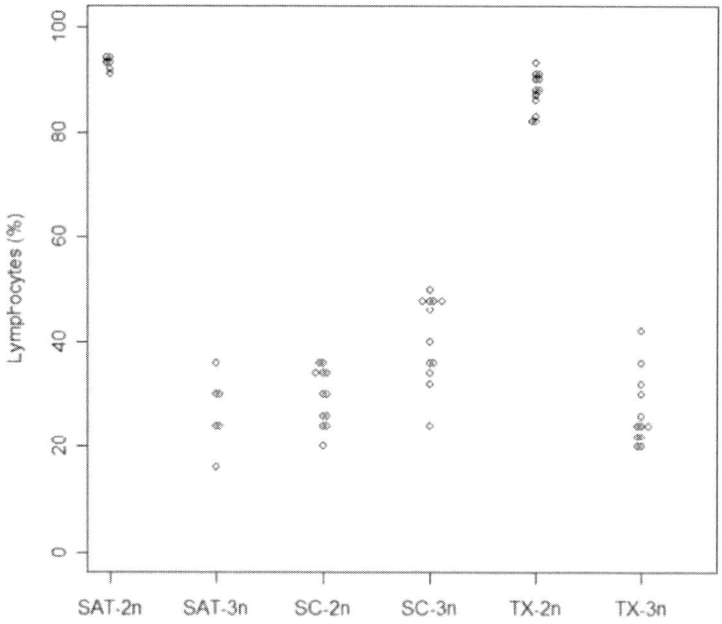

Figure 5. Lymphocytes.

Biochemistry Panel Values - Protein

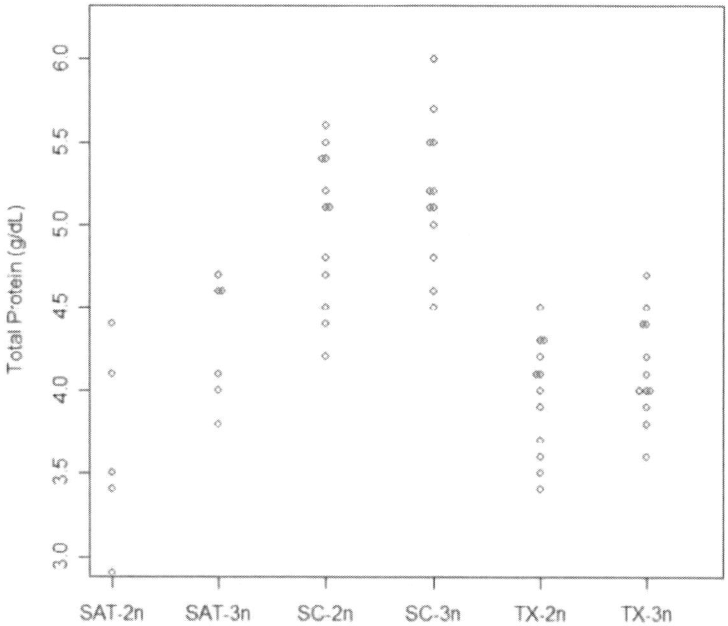

Figure 6. Total Protein.

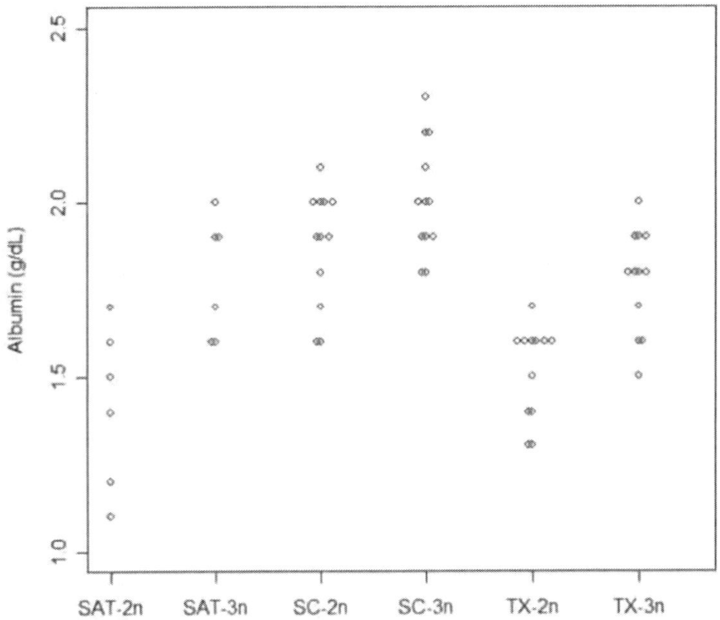

Figure 7. Albumin.

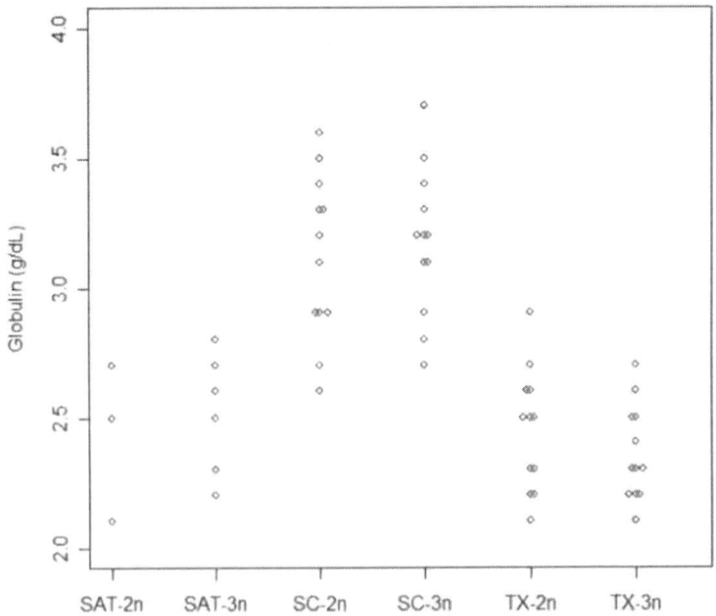

Figure 8. Globulin.

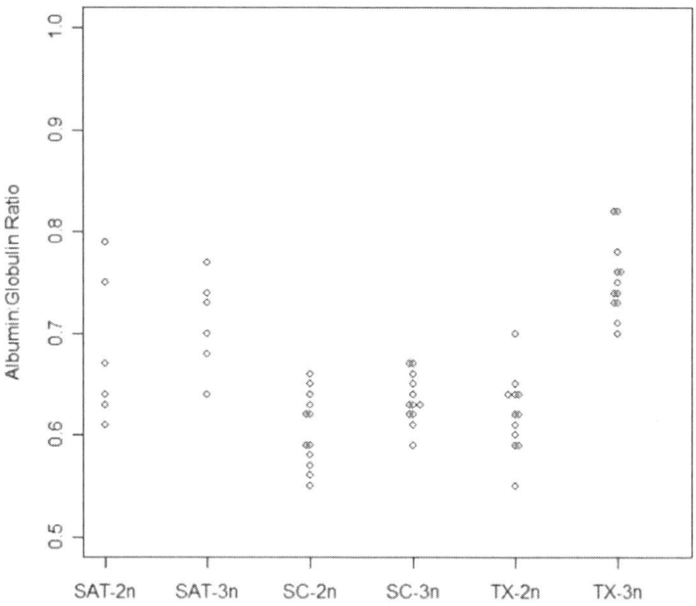

Figure 9. Alubmin:Globulin Ratio.

Biochemistry Panel Values – Electrolytes

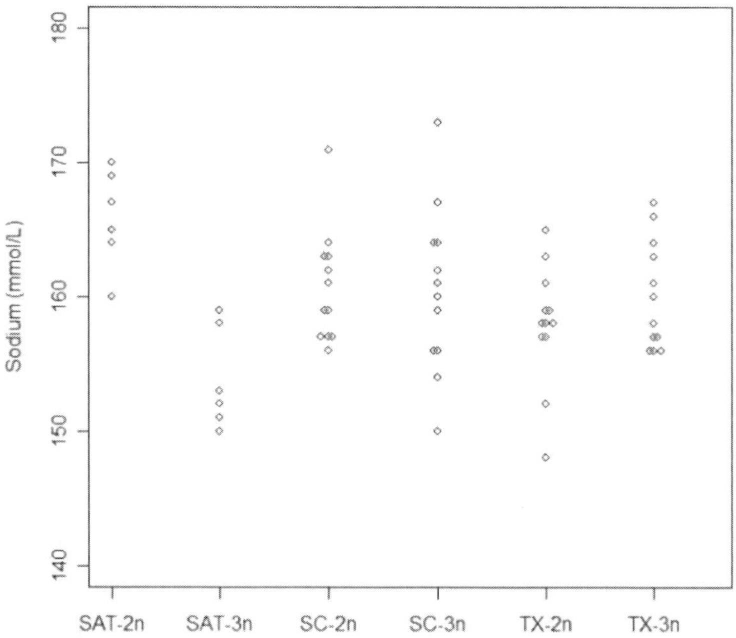

Figure 10. Sodium.

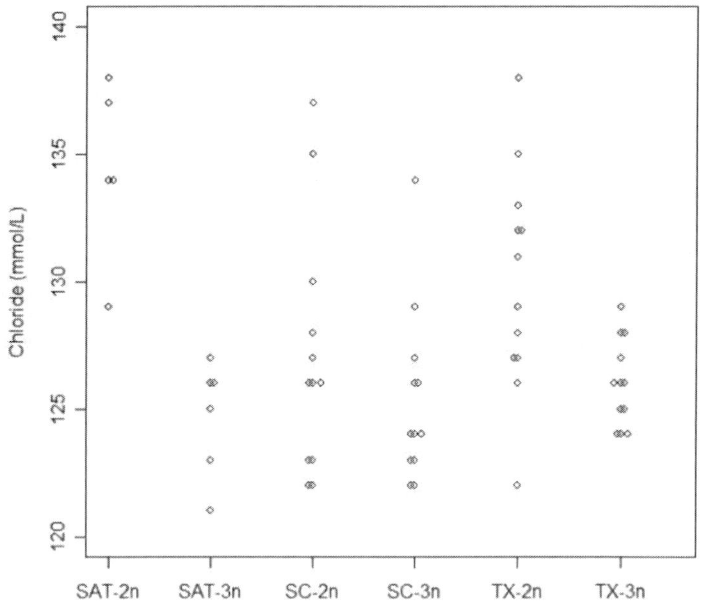

Figure 11. Chloride.

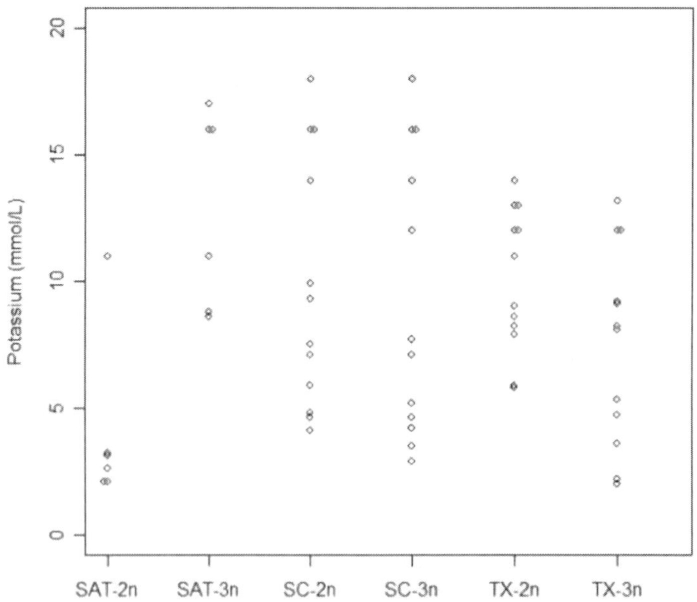

Figure 12. Potassium.

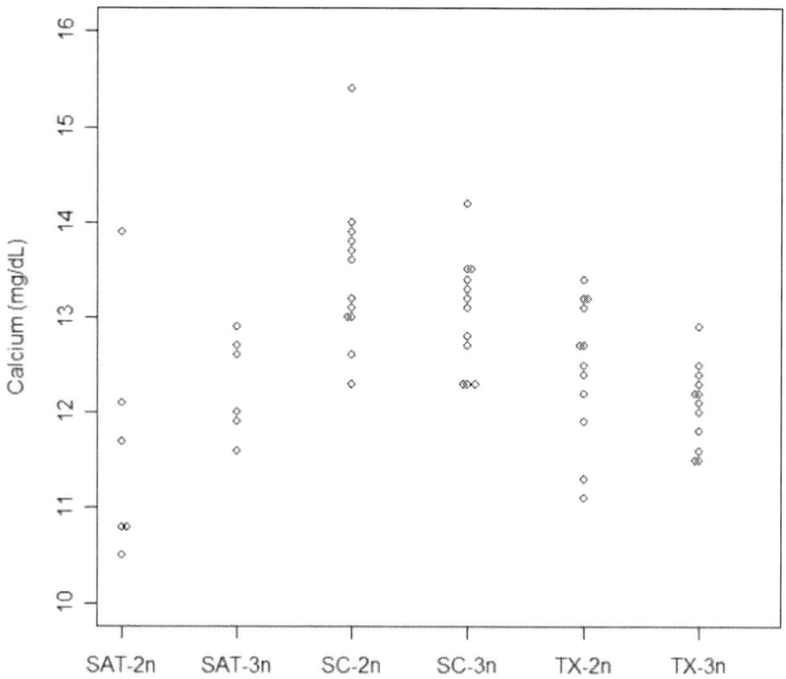

Figure 13. Calcium.

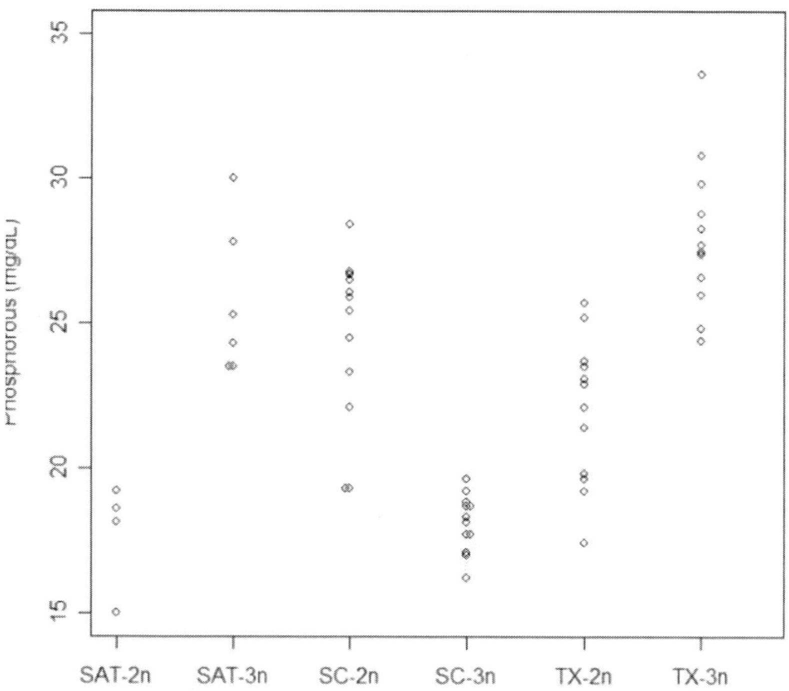

Figure 14. Phosphorous.

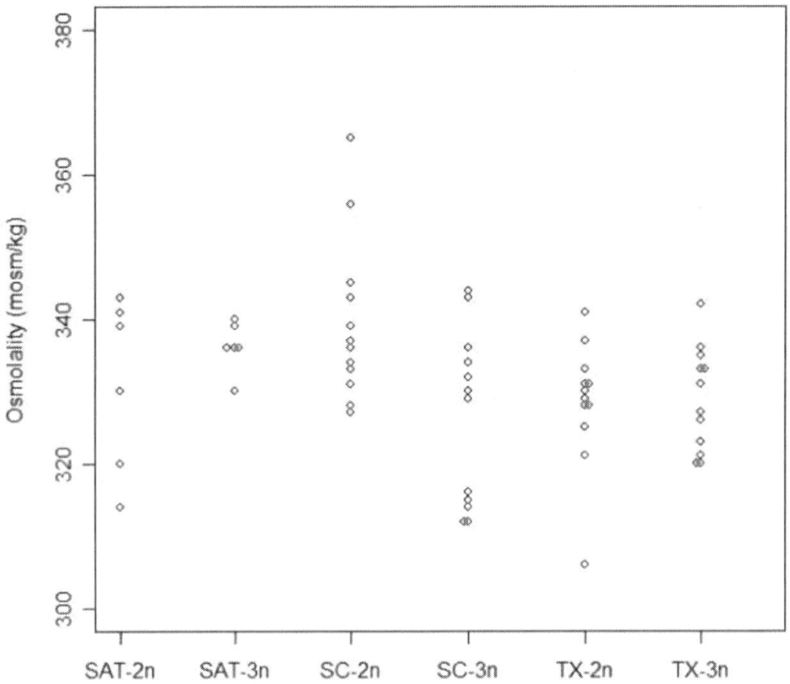

Figure 15. Osmolarity.

Biochemistry Panel Values – Enzymes and Metabolites

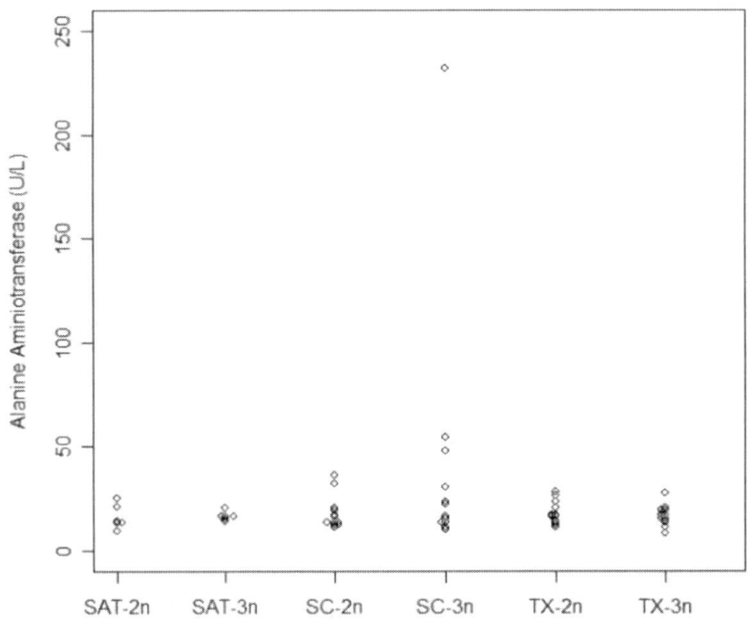

Figure 16. Alanine Aminotransferase.

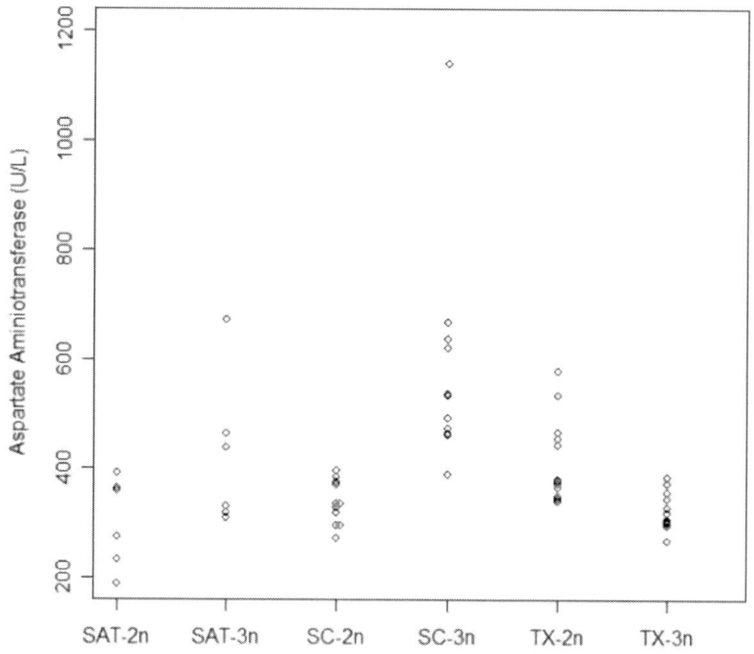

Figure 17. Aspartate Aminotransferase.

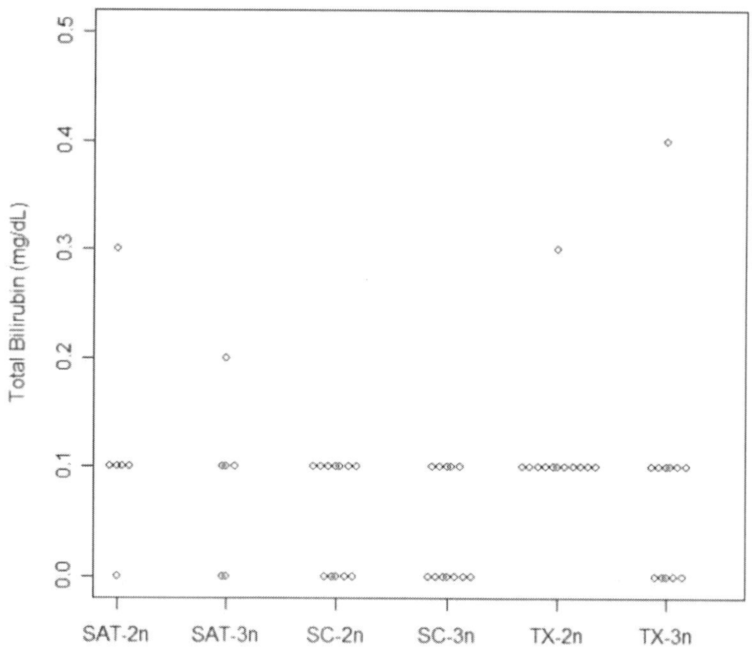

Figure 18. Total Bilirubin.

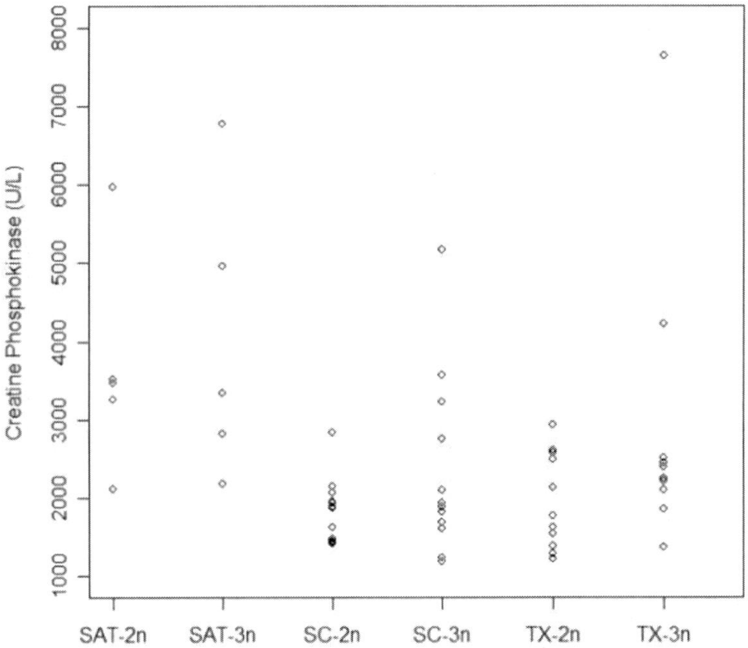

Figure 19. Creatine Phosphokinase.

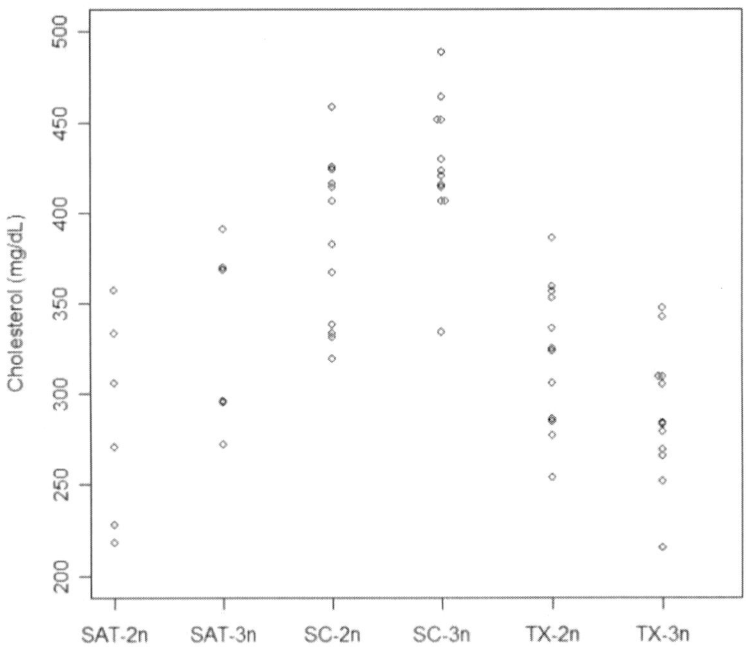

Figure 20. Cholesterol.

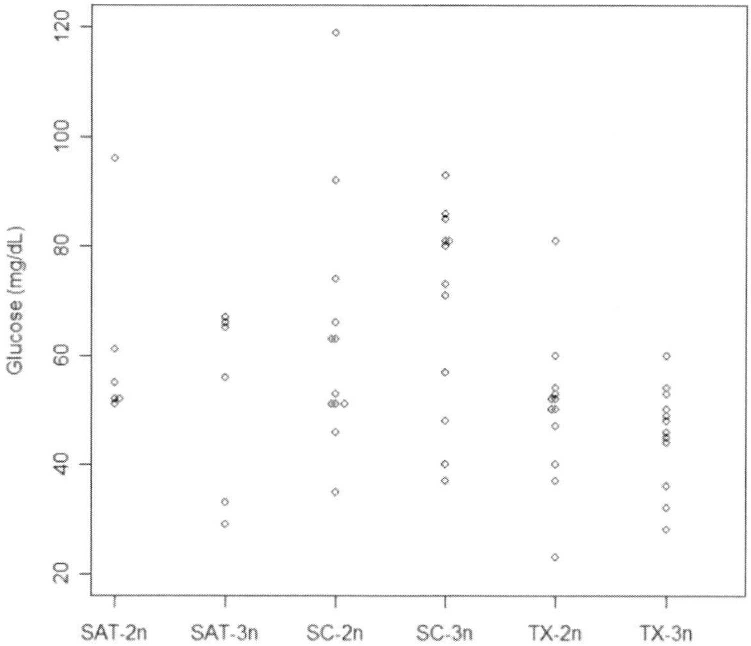

Figure 21. Glucose.

B. Food Safety Appendix A – Regulatory Method

FOI Summary for Aquadvantage Salmon Regulatory Method

AquAdvantage Salmon have an enhanced growth phenotype when compared to non-genetically engineered (non-GE) salmon which allows them to grow faster. However, at market size the AquAdvantage Salmon are not phenotypically distinguishable from non-GE fish. The FDA requires that animals of species that are traditionally consumed as food must have an analytical method designed to detect the presence of the AquAdvantage construct in tissues or edible products from these GE animals.

A multiplex polymerase chain reaction (PCR) procedure used by Aqua Bounty Technologies, Inc. (ABT) to confirm both the genotype of the AquAdvantage Salmon and the location of the integration site for the AquAdvantage construct at the α- locus was provided to FDA. This is the method ABT proposes to use for routine genotyping and surveillance of product durability and hence it is their proposed Regulatory Method. Some aspects of this multiplex PCR method were modified by FDA/CVM to achieve a robust, repeatable, accurate, molecular regulatory method.

The regulatory method is capable of discriminating between the AquAdvantage Salmon and their unmodified, non-GE counterparts. ABT's proposed PCR assay uses three primer pairs. The first primer set provides confirmation of the presence of the approved AquAdvantage construct in the fish. Samples from both AquAdvantage Salmon and non-GE fish will generate two DNA amplicons corresponding to the endogenous growth hormone gene. However, samples from AquAdvantage Salmon will also amplify an additional DNA fragment unique to the AquAdvantage construct. PCR assays using two additional primer pairs amplify DNA regions at the 5' and 3' junctions of the AquAdvantage construct as further evidence in support of the integration event in the EO-1α lineage. These two primer sets provide additional assurance about the genetic nature of the fish samples and help identify these fish as the AquAdvantage Salmon containing the AquAdvantage construct at the α- locus.

In addition to being able to distinguish AquAdvantage Salmon from non-GE fish in a mixed population this method is also capable of (1) identifying edible tissue from AquAdvantage Salmon, (2) identifying a durability failure, and (3) discriminating between the AquAdvantage Salmon and a knock-off (other GE fish containing similar constructs that are not AquAdvantage Salmon).

We conclude that this method is suitable for the purposes proposed by ABT and is capable of determining genotype and confirming rDNA integration at the described α- locus. Further, the FDA/CVM modified method has been peer-validated at the Agency's Office of Regulatory Affairs (ORA) district laboratory. This validated PCR method meets the Agency's requirements for a Regulatory method to identify the presence of the AquAdvantage construct in fish and is available from CVM, FDA, 7500 Standish Place, Rockville, MD 20855.

C. Food Safety Appendix B – Figures from the Regulatory Fish Encyclopedia Study

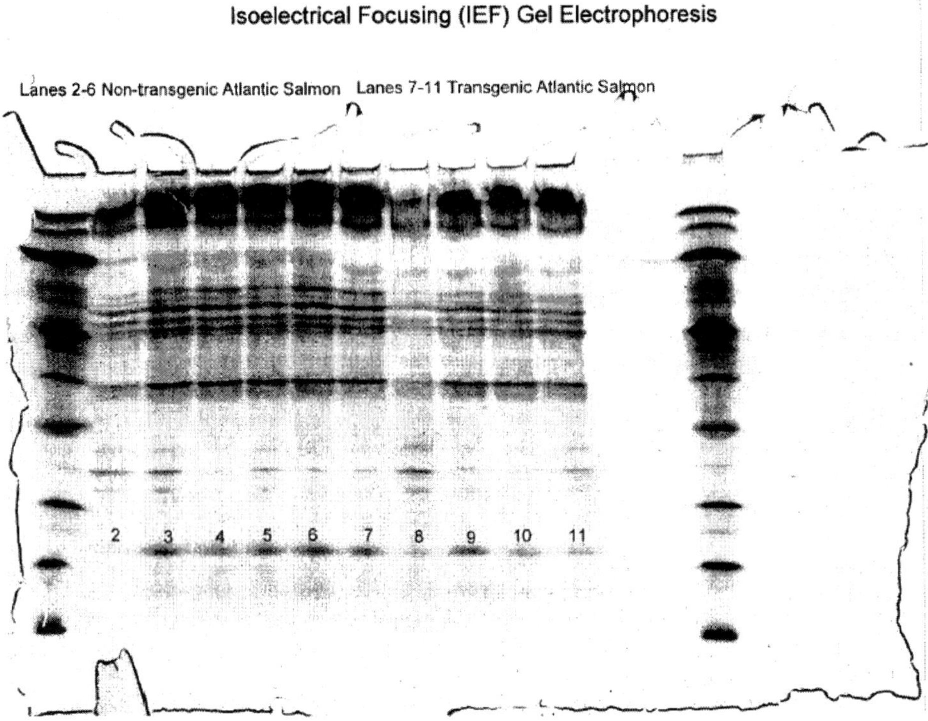

2-Dimensional Gel Electrophoresis

Non-transgenic Atlantic Salmon (Control)

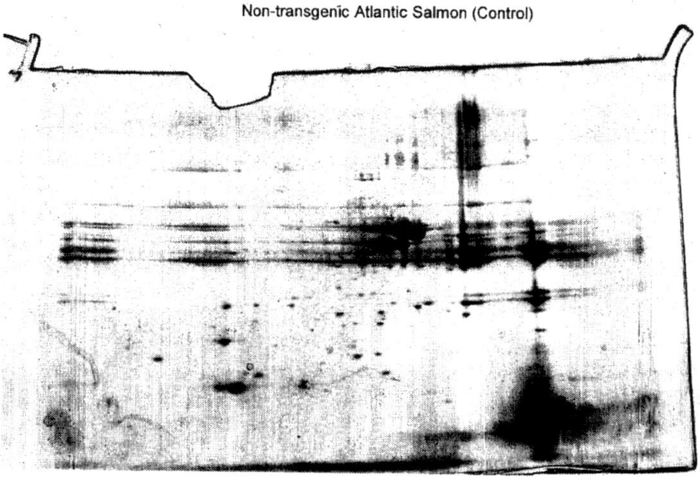

Transgenic Atlantic Salmon

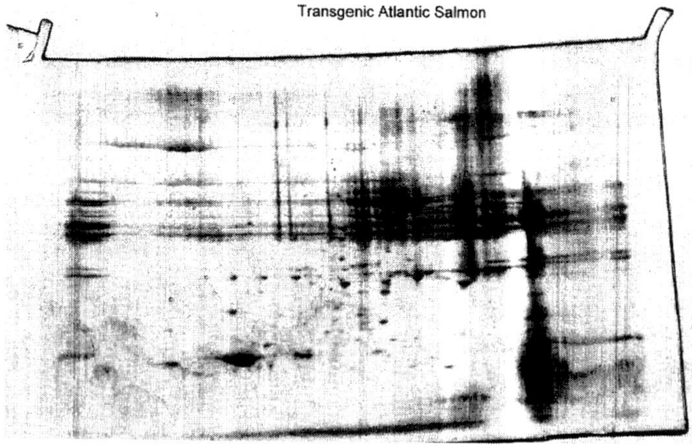

Store bought Atlantic Salmon (Reference Sample)

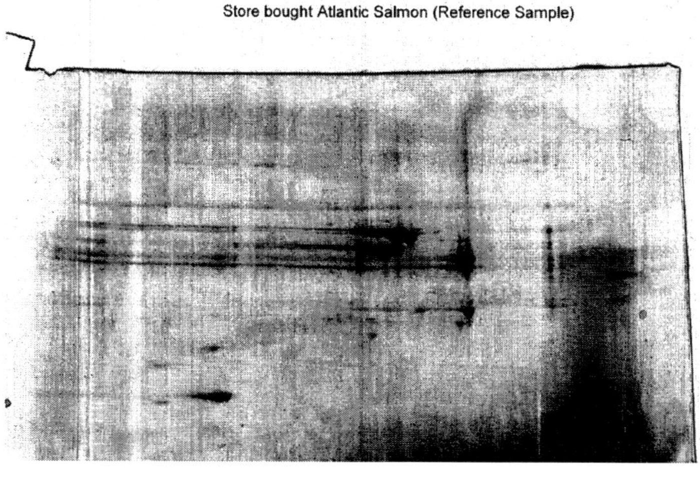

D. Food Safety Appendix C – Species Specificity of Growth Hormone

With the intent of determining the ability of growth hormone to bind to homologous growth hormone receptors and potentially activate them physiologically across vertebrate species we performed a survey of existing scientific literature. Based on the results of this search we conclude that Growth Hormone (GH) binding efficiency decreases as one compares up the phylogenetic tree, meaning that the GH of lower vertebrates do not bind the Growth Hormone Receptor (GHR) of higher vertebrates with sufficient affinity to activate the signaling cascade that causes the somatotropic effects associated with GH activity. Thus, fish GH would not activate human GHR and would therefore be physiologically ineffective.

Initial interest in the ability of cross species activity of GH resulted from an attempt to use porcine pituitary extracts against dwarfism in humans in the 1950s. These pituitary extracts were effective in animals but not in humans. Porcine GH shows *in vivo* cross species reactivity when administered subcutaneously or intramuscularly to rats (wild type and hypophysectomized), and dogs but not in hypophysectomized monkeys or human cells. GH derived from bovine, ovine, porcine and whale pituitaries were ineffective in humans[1,2]. Further, porcine pituitary extracts when used in humans did not stimulate growth, increase plasma free fatty acid levels, decrease plasma alpha-amino nitrogen, impair glucose tolerance or cause hyperinsulinaemia in GH deficient children[1].

In rats recombinant porcine GH (rpGH) was not orally active and showed no effect when orally administered and was not bioactive across the GI tract. rpGH did not bioaccumulate in the serum or cause an immune reaction. rpGH when administered at 4 mg/kg (considered a high dose in rats) by oral gavage caused no treatment related toxicity and the conclusion was that rpGH being a protein was subject to degradation by digestive enzymes in the gastrointestinal tract. Subcutaneous administration of pGH and rpGH in both wild type and hypophysectomized rats led to stimulation of growth[1]. Similarly, bovine GH is effective when administered subcutaneously to rats and has no significant effect via the oral route [2].

Injecting pGh or rpGH in pigs resulted in increased serum GH levels but not in physiologically significant increased muscle GH concentrations when measured at 27 hours post-treatment[1]. The peak serum GH levels after intramuscular injection of pGH were seen at 6 hours and the mean half life of pGH was 4 minutes (fast phase) and 38 minutes (slow phase). Serum GH levels in untreated pigs was in the range of 1.6 to 7 ng/ml[1].

Species specificity is related to several features of both the GH molecule and its cognate receptor. Most non-primate GH differ from each other by zero to four amino acids in the mature full length peptide of 190 amino acids[3] while human and rhesus GH differ from non-primate GH by 59-63 amino acids a difference of ~33% ([3] and see Table 1 below), indicating a large evolutionary shift in GH protein sequence with potential for significant secondary and tertiary structure effects that could affect function. Porcine GHs (both pGH and rpGH) are 66% similar to human GH they still do not bind sufficiently to human GHR as determined by their inability to displace bound ^{125}I-hGH *in vitro*[1]. Fish GH is 34% identical to human GH (see Table 1) and does not bind human GHR effectively enough to cause its activation. Bacterially expressed recombinant Dolphinfish GH had no mitogenic activity *in vitro* on cloned hGH receptor expressing cells. However, both recombinant common carp and recombinant dolphinfish GH stimulated rabbit GHR expressing cells[5]. Competitive binding assays with purified recombinant proteins and homologous GHRs provided the following results: IC_{50} dophinfish GH = 3.06 nM, gilthead seabream ~2 nM and K_a values of 0.77, 1.30

and 0.52 nM^{-1} for dolphinfish, common carp and gilthead seabream GH respectively[5]. Ruminant placental lactogen shows some *in vivo* effect in fish but is not as efficient a somatotrophic agent as homologous fish GHR[5]. Human GH has a high affinity for non-primate GHRs but non-primate GHs have ~3000X lower affinity for human GHR than human GH[3 ,4]. Essentially this results in an inability of non-primate GHs to stimulate human GHR[3].

The molecular basis for this species specificity is a change in human GH amino acid sequence at position 171 from histidine to aspartic acid (Asp). This Asp residue at position 171 is shared among all primates and therefore believed to have occurred in a common ancestor of simians. Tarsiers, lemurs and non-primates share a histidine at this position, along with all other vertebrates from cartilaginous fish to mammals, indicating that this is the ancestral form of the protein[3 ,4]. Additionally, human GHR has a Leucine (Leu) to Arginine (Arg) change at position 43 which causes causes an unfavorable charge repulsion/steric hindrance between Arg43 and non-primate GH His170 resulting in the decreased binding efficiency and inability for receptor activation[4]. The Asp at 171 of human GH forms a favorable salt bridge with Arg43 of human GHR allowing efficient binding and activation[3]. Thus, non-primate GHs have little to no binding and activation potential for human GHR, especially if ingested via the oral route.

Table 1. Percentage identity and similarity between derived amino acid sequences of Growth Hormone1 (GH1) coding regions of various vertebrate species

Identity / Similarity	Chinook GH1	Chinook GH2	Salmo GH1	Salmo GH2	Danio GH1	Gallus GH1	Mus GH1	Bos GH1	Human GH1
Chinook GH1		93.8% 97.1	95.2 99.0	94.3 98.1	62.3 79.2	37.7 57.4	36.1 53.9	35.7 52.7	34.4 50.9
Chinook GH2			96.7 97.6	97.6 98.6	63.2 79.7	37.7 55.6	34.7 52.0	35.7 50.4	33.5 50.0
Salmo GH1				97.1 98.1	63.2 79.7	38.6 57.4	36.1 53.9	36.6 52.2	33.9 50.0
Salmo GH2					62.9 79.3	37.7 56.1	36.1 53.9	36.6 51.8	33.9 50.0
Danio GH1						41.1 59.4	39.9 57.3	40.2 57.5	35.2 52.2
Gallus GH1							74.2 88.9	72.0 84.9	56.4 73.6
Mus GH1								82.9 88.9	66.5 78.0
Bos GH1									66.7 76.7
Human GH1									

Genbank Accession Numbers for the amino acid sequences used are listed below. All the sequences were downloaded from the NCBI database (http://www.ncbi.nlm.nih.gov/sites/gquery).

Homo sapiens GH1 Accession#: NM_000515.3

Bos taurus GH1 Accession#: NP_851339.1
Mus musculus GH1 Accession#: NP_032143.1
Gallus gallus GH1 Accession#: NP_989690.1
Danio rerio GH1 Accession#: NP_001018328.2
Salmo salar GH1 Accession#: AY614010.1
Salmo salar GH2 Accession#: EU621899.1
Onchorhynchus tshawytscha GH1 Accession#: EU621900.1
Onchorhynchus tshawytscha GH2 Accession#: EU621901.1

The AquAdvantage construct contains the Chinook salmon (*Onchorhynchus tshawytscha)* GH1 open reading frame including both the 188 amino acid coding region and the 22 amino acid signal peptide. Sequences were aligned using the European Bioinformatics Institute (EBI) online tool EMBOSS (http://www.ebi.ac.uk/Tools/emboss/align) which uses the Needleman-Wunsch algorithm to produce a full path matrix upon implementing a Pairwise Sequence Alignment. Our analysis used the default settings for protein alignments - a Blosum62 matrix and Gap penalties of 10.0 (Open) and 0.5 (Extend).

References

[1] Porcine somatotropin (WHO Food Additives Series 43). 52nd Meeting of the Joint FAO/WHO Expert Committee on Food Additives (JECFA). World Health Organization- IPCS, Geneva, 2000. www.inchem.org/documents/jecfa.jecmono/v43jec06.htm

[2] Juskevich, J. & Guyer, C. G. (1990). Bovine Growth Hormone: Human Food Safety Evaluation. *Science.*, *249*, 875-884.

[3] Liu, J. –C., Makova, K. D., Adkins, R. M., Gibson, S. & Li, W. –H. (2001). Episodic Evolution of Growth Hormone in Primates and Emergence of the Species Specificity of Human Growth Hormone Receptor. *Molecular Biology and Evolution.*, *18*(6), 945-953.

[4] Souza, S. C., Frick, G. P., Wang, X., Kopchick, J. J., Lobo, R. B. & Goodman, H.M. (1995). A Single arginine residue determines species specificity of the human growth hormone receptor. *Proc. Natl. acad. Sci. USA.*, *92*, 959-963.

[5] Paduel, A., Chapnik-Cohen, N., Gertler, A. & Elizur, A. (1999). Preparation and Characterization of Recombinant Dolphin Fish (Coryphaena hippurus) Growth Hormone). *Protein Expression and Purification.*, *16*, 417-423.

E. Food Safety Appendix D – Statistical Analysis Results for Composition

Results from the statistical analysis of selected analytes are provided below. Data were analyzed using analysis of variance with group (FC, SC, TX), ploidy (diploid, triploid) and the group-by- ploidy interaction (group*ploidy) included in the model as fixed effects. The analysis results are interpreted with the understanding that the estimated p-value may be under estimated (too small) because comparisons are generated after the data were examined. However, for exploratory analyses, this is an acceptable strategy.

If the group-by-ploidy interaction (group*ploidy) is considered significant, the TX group mean was compared to the FC and SC group means separately within ploidy. If the group-by-ploidy interaction was not considered significant, and the group effect was considered significant, the TX group mean was compared to the FC and SC group means without regard to ploidy.

Analyte	Effect	Probability*
calcium	group	0.1447
	ploidy	0.6262
	group*ploidy	0.8451
copper	group	0.6270
	ploidy	0.8828
	group*ploidy	0.7579
manganese	group	0.7421
	ploidy	0.3552
	group*ploidy	0.4949
potassium	group	0.0001
	ploidy	0.6698
	group*ploidy	0.1314
serine	group	0.0016
	ploidy	0.4425
	group*ploidy	0.1224
vitamin b6	group	0.0002
	ploidy	0.0442
	group*ploidy	0.1697
zinc	group	0.1094
	ploidy	0.5896
	group*ploidy	0.4116
folic acid	group	0.0756
	ploidy	0.0195
	group*ploidy	0.0277
iron	group	0.6575
	ploidy	0.3233
	group*ploidy	0.3109
magnesium	group	<0.0001
	ploidy	0.0466
	group*ploidy	0.0065
niacin	group	<0.0001
	ploidy	0.0977
	group*ploidy	0.0223
phosphorous	group	0.0039
	ploidy	0.4593
	group*ploidy	0.0489

* Probability associated with the F-test for the specified effect.

Analyte	Group*	Least Squares Mean	Standard Error	P-value**
calcium	FC	31.490	1.9787	
	SC	30.057	1.0937	
	TX	27.573	1.1424	
copper	FC	0.064	0.0110	
	SC	0.069	0.0061	
	TX	0.075	0.0063	
manganese	FC	0.028	0.0099	
	SC	0.033	0.0055	
	TX	0.027	0.0057	
potassium	FC	375.500	6.8889	0.3911
	SC	394.244	3.8080	<0.0001
	TX	368.633	3.9773	
serine	FC	0.761	0.0196	0.7779
	SC	0.811	0.0108	0.0006
	TX	0.755	0.0113	
vitamin b6	FC	8.002	0.2691	0.0001
	SC	8.736	0.1488	0.0086
	TX	9.318	0.1554	
zinc	FC	0.568	0.0243	
	SC	0.515	0.0135	
	TX	0.509	0.0141	

* Group: FC = farm control, SC = sponsor control, TX = transgenic.

** The P-value associated with comparing the TX mean to the FC or SC mean.

Analyte	Ploidy	Group*	Least Squares Mean	Standard Error	P-value**
folic acid	diploid	FC	0.366	0.0397	0.0009
		SC	0.272	0.0209	0.0391
		TX	0.207	0.0229	
	triploid	FC	0.212	0.0397	0.7145
		SC	0.231	0.0229	0.9589
		TX	0.229	0.0229	
iron	diploid	FC	0.65	0.088	0.1159
		SC	0.46	0.046	0.7180
		TX	0.49	0.051	
	triploid	FC	0.39	0.088	0.1065
		SC	0.49	0.051	0.3965
		TX	0.55	0.051	
magnesium	diploid	FC	26.100	0.7244	0.6799
		SC	26.706	0.3818	0.0973
		TX	25.753	0.4182	
	triploid	FC	25.020	0.7244	0.0988
		SC	27.260	0.4182	<0.0001
		TX	23.620	0.4182	

Analyte	Ploidy	Group*	Least Squares Mean	Standard Error	P-value**
niacin	diploid	FC	86.280	3.3301	0.0001
		SC	91.033	1.7551	<0.0001
		TX	102.527	1.9226	
	triploid	FC	91.500	3.3301	0.8157
		SC	85.813	1.9226	0.0181
		TX	92.400	1.9226	
phosphorous	diploid	FC	263.000	6.2586	0.8686
		SC	265.167	3.2986	0.4938
		TX	261.800	3.6134	
	triploid	FC	258.400	6.2586	0.3139
		SC	272.067	3.6134	0.0001
		TX	251.067	3.6134	

Analyte	Effect	Probability*
eicosdienoic acid	group	<0.0001
	ploidy	0.6500
	group*ploidy	0.2720
linoleic acid	group	0.0002
	ploidy	0.9317
	group*ploidy	0.3109
linolenic acid	group	0.0001
	ploidy	0.5104
	group*ploidy	0.5594
oleic acid	group	<0.0001
	ploidy	0.3675
	group*ploidy	0.1397
arachidic acid	group	<0.0001
	ploidy	0.6600
	group*ploidy	0.0099
docosahexaenoic acid	group	<0.0001
	ploidy	0.5783
	group*ploidy	0.0073
docosapentaenoic acid	group	<0.0001
	ploidy	0.9796
	group*ploidy	0.0968
eicosopentaenoic acid	group	<0.0001
	ploidy	0.5967
	group*ploidy	0.0241
free fatty acids	group	0.0045
	ploidy	0.7525
	group*ploidy	0.1044
palmitoleic acid	group	<0.0001
	ploidy	0.5604

(Continued)

Analyte	Effect	Probability*
	group*ploidy	0.0296
palmitic acid	group	<0.001
	ploidy	0.5154
	group*ploidy	0.0032
stearic acid	group	<0.0001
	ploidy	0.6767
	group*ploidy	0.0097
total fatty acids	group	<0.0001
	ploidy	0.5646
	group*ploidy	0.0207

Analyte	Group*	Least Squares Mean	Standard Error	P-value**
eicosdienoic acid	FC	0.05	0.005	0.2852
	SC	0.04	0.003	<0.0001
	TX	0.06	0.003	
linoleic acid	FC	0.67	0.068	0.3403
	SC	0.51	0.038	<0.0001
	TX	0.74	0.039	
linolenic acid	FC	0.18	0.028	0.1004
	SC	0.13	0.016	<0.0001
	TX	0.23	0.016	
oleic acid	FC	2.88	0.234	0.1245
	SC	2.00	0.129	<0.0001
	TX	3.30	0.135	

Analyte	Ploidy	Group*	Least Squares Mean	Standard Error	P-value**
arachidic acid	diploid	FC	0.03	0.003	0.0916
		SC	0.01	0.001	0.0014
		TX	0.02	0.002	
	triploid	FC	0.02	0.003	0.2258
		SC	0.01	0.002	<0.0001
		TX	0.03	0.002	
docosahexaenoic acid	diploid	FC	1.44	0.116	0.3061
		SC	1.04	0.061	0.0065
		TX	1.30	0.067	
	triploid	FC	1.48	0.116	0.6451
		SC	0.86	0.067	<0.0001
		TX	1.55	0.067	

Analyte	Ploidy	Group*	Least Squares Mean	Standard Error	P-value**
docosapentaenoic acid	diploid	FC	0.41	0.052	0.2325
		SC	0.30	0.027	<0.0001
		TX	0.48	0.030	
	triploid	FC	0.46	0.052	0.3464
		SC	0.22	0.030	<0.0001
		TX	0.52	0.030	
eicosopentaenoic acid	diploid	FC	1.13	0.116	0.3192
		SC	0.67	0.061	0.0006
		TX	1.00	0.067	
	triploid	FC	1.22	0.116	0.8663
		SC	0.50	0.067	<0.0001
		TX	1.19	0.067	
free fatty	diploid	FC	0.05	0.014	0.3447
		SC	0.07	0.007	0.9515
		TX	0.07	0.008	
	triploid	FC	0.03	0.014	0.0002
		SC	0.07	0.008	0.0655
		TX	0.09	0.008	
palmitoleic acid	diploid	FC	0.97	0.086	0.0952
		SC	0.61	0.046	0.0051
		TX	0.80	0.050	
	triploid	FC	0.98	0.086	0.9575
		SC	0.51	0.050	<0.0001
		TX	0.97	0.050	
palmitic acid	diploid	FC	2.11	0.173	0.0118
		SC	1.16	0.091	0.0021
		TX	1.59	0.100	
	triploid	FC	1.71	0.173	0.1752
		SC	0.96	0.100	<0.0001
		TX	1.99	0.100	
stearic acid	diploid	FC	0.43	0.041	0.2595
		SC	0.26	0.022	0.0008
		TX	0.37	0.024	
	triploid	FC	0.36	0.041	0.0367
		SC	0.21	0.024	<0.0001
		TX	0.46	0.024	
total fatty acid	diploid	FC	15.20	1.283	0.1483
		SC	9.76	0.676	0.0017
		TX	13.03	0.741	
	triploid	FC	15.14	1.283	0.6574
		SC	8.39	0.741	<0.0001
		TX	15.80	0.741	

F. Food Safety Appendix E – Vitamin B6 – Explanation of Conversion Factor

In this review, we report vitamin B6 concentrations in free base form. ABT used a standard yeast-based biochemical assay to measure the protonated ('salt') form of vitamin B6 (pyroxidine HCl). In order to be able to compare vitamin B_6 levels across studies, a conversion was required. We consulted with Darryl Sullivan of Covance Inc. on the conversion issue.

Vitamin B6 in peer-reviewed scientific literature is usually reported as the free base form (pyroxidine) and measured via HPLC. Conversion from salt to free base is accomplished by multiplying the salt form of Vitamin B6 by a coefficient of 0.823. This calculation accounts for the removal of HCl from the molecule[31]. We therefore refer to vitamin B6 in its free base form throughout this review.

Environmental Assessment for AquAdvantage Salmon

http://www.fda.gov/downloads/AdvisoryCommittees/CommitteesMeetingMaterials/VeterinaryMedicineAdvisoryCommittee/UCM224760.pdf

End Notes

[1] Throughout this document the terms "FDA," "CVM," or "agency" are used to denote formal agency action or policy; implementing statutory authority, regulations; and issuing or responding to recommendations in Guidance for Industry documents. The terms "we" and "us" refer to CVM staff when evaluating data and information.

[2] "ABT salmon" is used throughout this document to refer to fish under ABT's control, though not necessarily triploid, hemizygous, female fish.

[3] Throughout this section we use the term "comparator" to refer to non-GE fish of a similar, but not identical, genetic background as AquAdvantage Salmon, including both diploid and triploid fish. These fish are appropriate controls for these studies given the inherent constraints of a selective breeding program.

[4] Once they reach smolt size, Atlantic salmon are normally transferred to seawater and reared to market size in open water net pens, however, the entire lifecycle of AquAdvantage Salmon occurs in freshwater.

[5] Form FDA 483 is used to communicate investigational observations that may need correction.

[6] Two fish were accidentally frozen at PEI and therefore were not suitable samples for histopathology analysis. One of the 23 sent for analysis was delayed; by the time the fish was investigated the cells had already deteriorated to the point that histopathological analysis was impossible.

[7] The design of the previous tanks used for rearing of early life stages supposedly did not allow for a thorough cleaning resulting in a chronic fungal problem.

[8] During the inspection of the facility by FDA/CVM, health records were verified by following a chain of custody for selected samples and auditing the original observations of the investigators. In addition, records were collected from various original log books and verified for accuracy against records submitted as part of ABT's application. In all instances, original observations were verified and data capture forms matched those submitted to FDA.

[9] Unless stated otherwise in the reference, these salmon are assumed to be diploid.

[10] These were GH transgenic Atlantic salmon produced by ABT, however, from the information provided it is impossible to determine whether or not they were in fact early generations of what is currently known as AquAdvantage Salmon.

[11] Oxygen solubility in water is inversely proportional to water temperature. Salmonids in general have higher oxygen requirements than most other fish and thus require lower water temperatures so that oxygen levels are not limiting.

[12] Benfey, T.J., (1999), The Physiology and Behavior of Triploid Fishes, Reviews in Fishery Science, Vol 7(1): 39-67.

[13] The 10% exceedance was chosen as an arbitrary value that triggers additional investigation to determine whether the exceedance has any biological significance; it does not imply that beyond a 10% difference there is an *a priori* safety concern.

[14] In the evaluation of the safety of bovine somatotropins, the 41st Joint Food and Agriculture Organization of the United Nations/ World Health Organization Expert Committee on Food Additives (JECFA) considered the daily production of IGF1 in different age groups, and estimated by calculation the total serum burden for a 15 kg child (50,000 ng), a 60 kg adult (714,000 ng) and a 50 kg teenager (1,220,000 ng), considering the mean IGF1 concentration in plasma, and assuming blood volume to be 5% of body weight.

[15] Federal Register Vol. 44, No. 62, p. 18722-18724, March 29, 1979.

[16] The global commons comprises those parts of the earth beyond national boundaries, principally the open ocean and living resources therein, and those parts held in common, such as the atmosphere.

[17] The NRC committee acknowledged that it is possible for risk to occur without establishment; however, this was considered a special case that could be addressed as such.

[18] Reference herein to specific sections, figures, and tables of the EA is for the public release version of the EA unless otherwise noted.

[19] The term "barriers" was used in the Performance Standards when discussing similar containment measures. The term includes physical or chemical barriers, mechanical barriers and biological barriers.

[20] Form FDA 483 is used to communicate investigational observations that may need correction.

[21] The inspection report reported a minimum of 2 forms of mechanical containment, but counted the primary and secondary screens in the effluent containment sump as only one form. Here these two stainless steel screens are considered to be independent forms of containment as they are physically distinct.

[22] For comparison, the salinity of ocean water typically ranges from 28 to 32 ppt, while freshwater has a salinity of less than 1 ppt.

[23] The upper incipient lethal level is the highest temperature that can be survived up to seven days.

[24] http://www.nodc.noaa.gov/cgi-bin/OC5/WOA09F/woa09f.pl?navigation=t_0_16_1_forward; last visited on August 26, 2010.

[25] These were GH transgenic Atlantic salmon produced by ABT; however, from the information provided in the articles it is impossible to determine whether or not they were in fact early generations of what is currently known as AquAdvantage Salmon.

[26] Genetic (XX) females that have been treated with an androgen (17-methyl testosterone) during early development so they produce milt and have the other sexual characteristics of a male fish. Crossing of milt from neomales with eggs from true females can produce only genetically female offspring.

[27] This scenario would only be applicable in instances where there are fertile, female, wild Atlantic salmon or conspecifics present, a very unlikely occurrence.

[28] The process of gynogenesis involves the destruction of the genetic component in fish sperm, use of those "empty" sperm for egg activation, and restoration of a diploid state in the activated egg by forced retention of the second polar body.

[29] Key:
SAT-2n = diploid non-GE age matched comparators; SAT-3n = triploid non-GE age matched comparators; SC-2n = diploid non- GE size matched comparators; SC-3n = triploid non-GE size matched comparators; TX-2n = diploid GE; TX-3n = triploid GE

[30] Note that the monocytes counts were zero (0) for all samples. Data are not presented graphically here as the figure would appear blank for all groups.

[31] Pyroxidine HCl ($C_8H_{11}NO_3$-HCl) molecular weight is equal to 205.64 g/mol. Free base pyroxidine ($C_8H_{11}NO_3$) has a molecular weight of 169.24 g/mol. The difference between the two forms is simply calculated as 205.64/169.34 = 0.823.

Index

C

D

E

F

G

H

I

J

K

L

M

Q

R

S

T

U

V

W

Y

Z